North American Shortwave Frequency Guide

Written by

Captain James D. Pickard,
Retired M.I.
KG7BG

Introduction & Maps & Contributions by

Bill Smith, N6MQS

artsci

Artsci, Inc.
P.O. Box 1428
Burbank, CA 91507 U.S.A.
(818) 843-4080
(818) 846-2298 FAX

North American Shortwave Frequency Guide

NUMBER 3 10 9 8 7 6 5 4 3 2 1

ISBN 0-917963-09-1 $19.95

Notice of Liability

artsci inc, books are available for bulk sales at quantity discounts. For information, please contact Marketing Manager, artsci inc, P.O. Box 1428, Burbank, CA 91507 FAX: (818) 846-2298

—Dedication—

This book is dedicated to:

Leslie, Brad & Stefanie

Printed in the United States of America

Contents

Maps & Country Guides

Frequency Listings

Introduction

We are again proud to present this updated version of our popular Shortwave Directory. In this version we have updated many pages of new shortwave programming, maps of the world and program listings by country.

Shortwave listening is becoming more popular day after day. The equipment is getting better and easier to use. There has never been a better time to turn on and tune in.

News and events are happening around the world 24 hours a day, every day. Thousands of shortwave radio stations are broadcasting news of these events from the four corners of the world. All it takes is a shortwave radio and antenna to tune in to these broadcasts.

Your local TV or radio station may let you know about some world event, if they consider it important and other local events allow the time. If the event is announced, it is often a biased report.

Music programs are also broadcasted. The selection varies from Rock music in other languages, to classical in its original language and ethnic music. Yes, you could also hear RAP music (even when broadcast from another country the leading "C" is removed).

Political discussions are transmitted from all over the globe. You'll be surprised at the point of views expressed. People from around the world have very different ideas about us and the meaning of world events.

If you are technically inclined, you can connect a computer or printer to your shortwave radio and receive weather maps, teletype news, personal communications and FAX pictures. Most of these broadcasts are sent 24 hours a day.

This guide presents the broadcasts in order by frequency. This format is the most useful when you are tuning around the shortwave bands and discover a broadcast. Look up the frequency on the dial and you can quickly find it in the guide.

Many countries' broadcasts are simultaneously transmitted on different frequencies. See the section on propagation. Finding a certain country's broadcast is also simple. Scanning the list of frequencies for a certain broadcast is simplified once you narrow down the list by understanding propagation of radio waves.

This book can be used in any country around the world. The broadcasts propagate world wide. The broadcast times are presented using the world time standard. See the section on Universal Time Coordinated.

You can greatly improve the number of broadcasts you can receive and the quality of the signals by using an outside antenna. Construction plans for some inexpensive and easy to build antennas are provided in the section "Shortwave Antennas". Nothing you do will have a greater impact on the quality of reception than a good antenna.

Many radio broadcasters will send a certificate of reception to you if you drop them a note telling them your name, address and the frequency and time of day you heard their broadcast. They are very interested in who is listening and where their signals are received.

You may, after tuning in the world, get bit by the bug and want to get further into the hobby. The next step is becoming a HAM or amateur radio operator. Becoming a HAM is very simple. The FCC has relaxed the testing requirements to become a HAM. Morse code is no longer required.

Simple multiple choice tests are given monthly by volunteer groups around the country. Books are available from $6.00 - $20.00 that contain all the possible questions and all the answers. Take a few nights and study the questions and you'll be ready to pass the test. Some groups have classes to help you pass the test.

Once you have passed the test and received your license, you can communicate with other amateurs around the world. You may even be able to use the same antenna you build for shortwave reception. There are many benefits to being an amateur operator. Visit your local electronic store and ask them how to contact a local amateur group.

Your comments about this book are always welcome. Write or FAX your comments to us. If we implement your suggestion, we may send free copies of our books to say thank you.

Universal Time Coordinated

What time does the program start?

Listening in on radio transmissions from around the world is very easy. In fact its almost as simple as watching a television program. With the help of your TV guide, you look up a program and find the time it is playing and on which channel.

Finding a shortwave program is almost as simple. Converting the program times is the only task. The world is a big place, that has 24 different time zones. Programs that can be heard around the world can be heard in any one of the 24 time zones. To resolve the time zone dilemma, all broadcasting around the world is listed using only one time zone.

The selection of a "Universal" time zone was selected years ago. It was originally known as Greenwich Mean Time. The local time zone of Greenwich, England is the "Universal" time zone. Today, Greenwich Mean time has been renamed Universal Time Coordinated (U.T.C.).

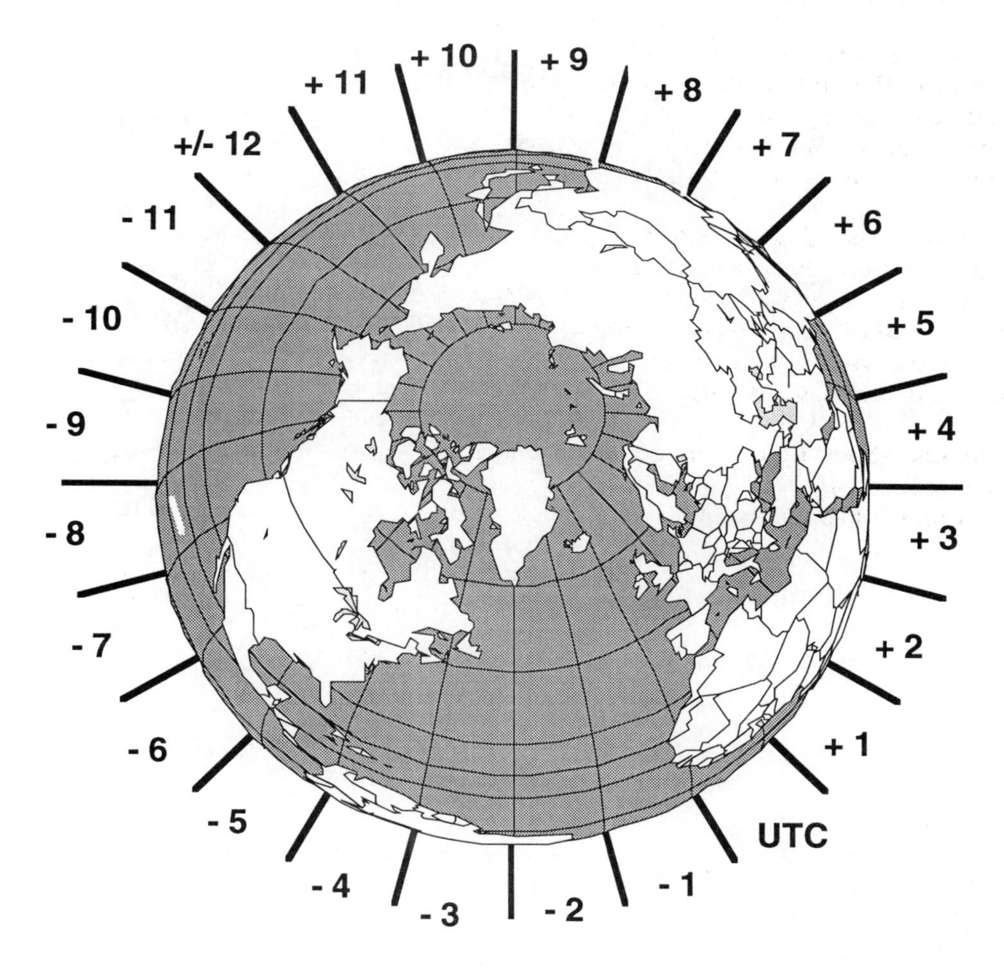

The world map presented above shows all 24 time zones and their relationship to U.T.C. Perhaps the easiest method of understanding how the time zones work is to imagine the sun moving around the earth. On the world map example the sun will move in a clockwise direction.

Place the sun over Greenwich, England, this would make it 12:00 noon U.T.C. If you move one time zone clockwise, subtract one hour. 12:00 U.T.C. is 11:00 local time. The time zone in California (PST) is minus (-) 8 hours U.T.C., or 4 am.

Now that you have a basic understanding of how U.T.C. works. Lets take an real example of a shortwave program and figure out what when you can hear it in your time zone

Example: Voice of America
9.715 MHz
0400-0600 UTC

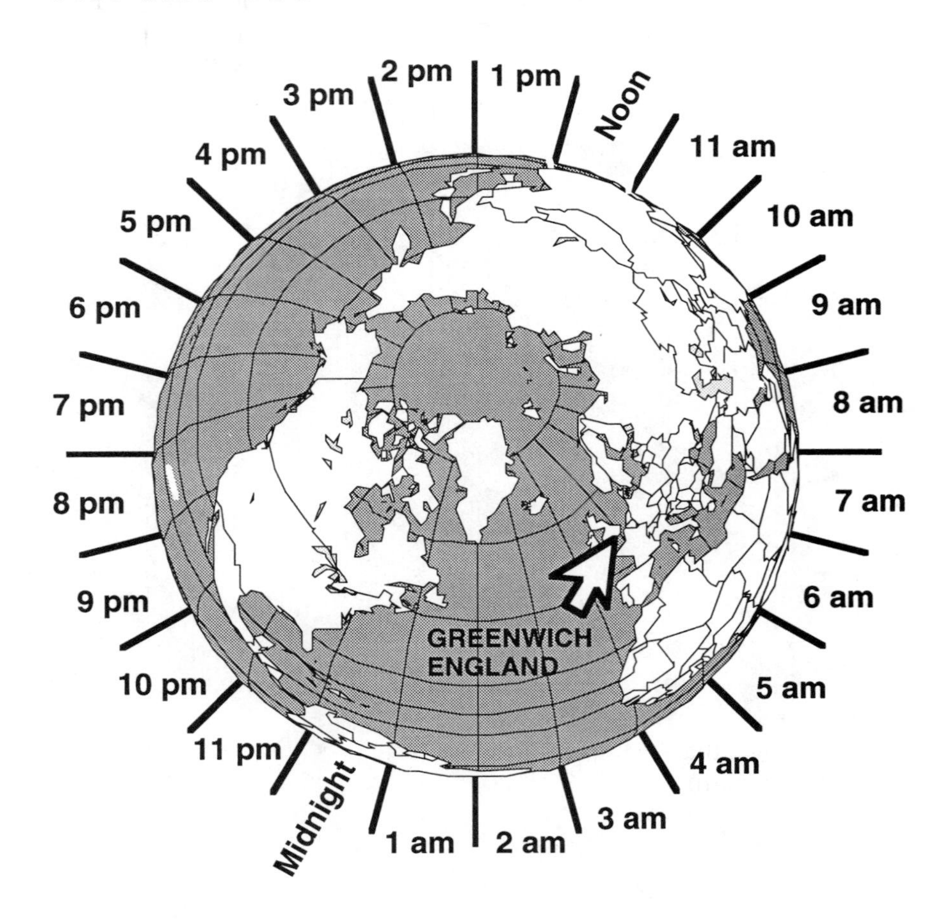

The program starts at 0400 or 4 am. The world map below places 4 am over Greenwich, England. As you move position around the globe, the start time is shown for each time zone.

If you live somewhere in North America, the table presented below will help you quickly compute U.T.C. time to your local time.

TIME CONVERSION CHART					
U.T.C.	PST	PDST MST	MDST CST	CDST EST	EDST
0:00	4 pm	5 pm	6 pm	7 pm	8 pm
1:00	5 pm	6 pm	7 pm	8 pm	9 pm
2:00	6 pm	7 pm	8 pm	9 pm	10 pm
3:00	7 pm	8 pm	9 pm	10 pm	11 pm
4:00	8 pm	9 pm	10 pm	11 pm	Midnight
5:00	9 pm	10 pm	11 pm	Midnight	1 am
6:00	10 pm	11 pm	Midnight	1 am	2 am
7:00	11 pm	Midnight	1 am	2 am	3 am
8:00	Midnight	1 am	2 am	3 am	4 am
9:00	1 am	2 am	3 am	4 am	5 am
10:00	2 am	3 am	4 am	5 am	6 am
11:00	3 am	4 am	5 am	6 am	7 am
12:00	4 am	5 am	6 am	7 am	8 am
13:00	5 am	6 am	7 am	8 am	9 am
14:00	6 am	7 am	8 am	9 am	10 am
15:00	7 am	8 am	9 am	10 am	11 am
16:00	8 am	9 am	10 am	11 am	Noon
17:00	9 am	10 am	11 am	Noon	1 pm
18:00	10 am	11 am	Noon	1 pm	2 pm
19:00	11 am	Noon	1 pm	2 pm	3 pm
20:00	Noon	1 pm	2 pm	3 pm	4 pm
21:00	1 pm	2 pm	3 pm	4 pm	5 pm
22:00	2 pm	3 pm	4 pm	5 pm	6 pm
23:00	3 pm	4 pm	5 pm	6 pm	7 pm

_ST - Standard Time
_DST - Daylight Savings Time

Now that you understand how world time zones work, its time to listen in on the broadcasts from around the world.

Shortwave Antennas

A hole in the sky

The most important component of a shortwave radio is the antenna. You could have purchased the most expensive shortwave receiver available (approximately $5,000) and your next door neighbor who purchased the budget model may receive 10 times more stations. The type and placement of an antenna is more important that the cost of the radio.

You will be attempting to listen to radio stations from around the world. The signals from these stations are commonly very weak by the time they travel the great distance to your home. Attempting to capture these signals with a piece of wire or telescoping antenna inside your home may turn out to be a very frustrating experience.

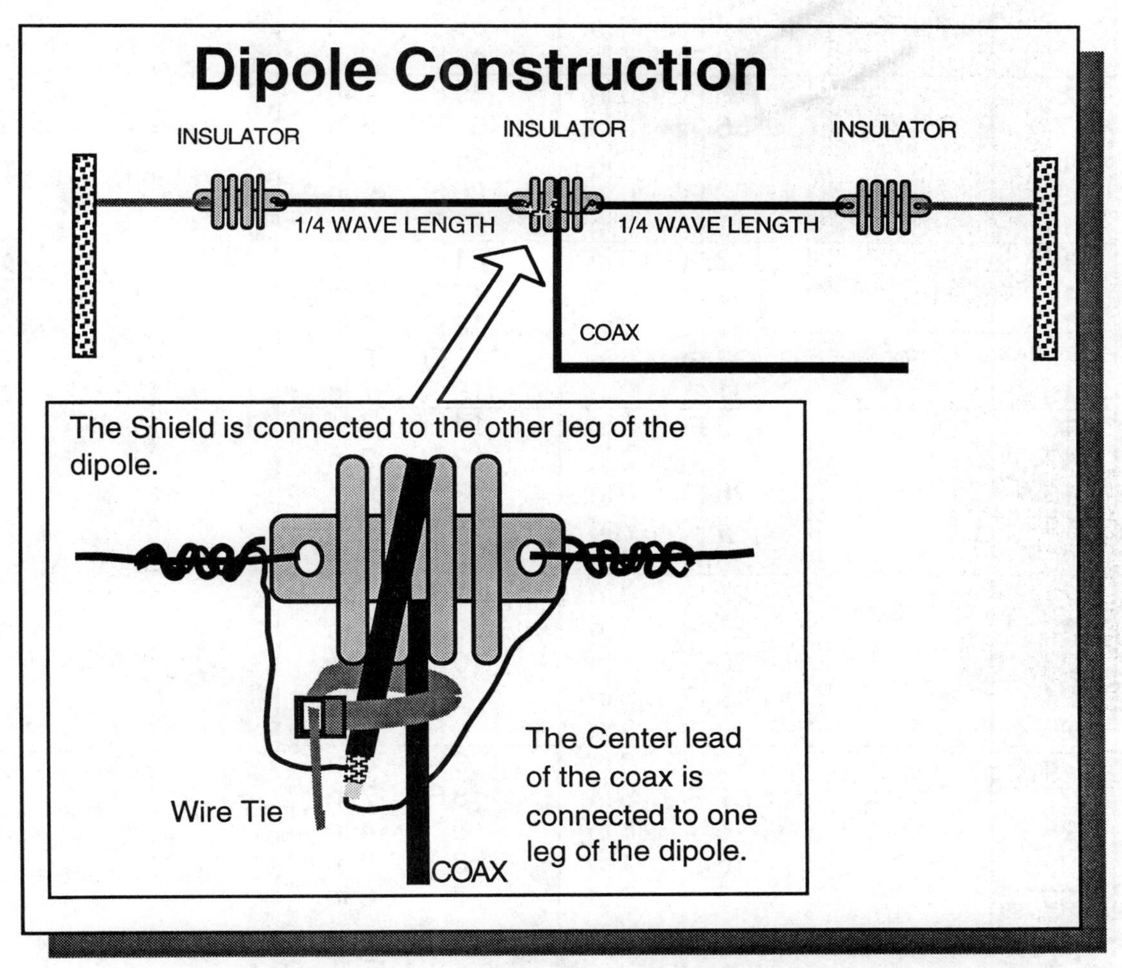

One of the easiest and most successful antennas to build is the Dipole.

The constructions plans above show the few parts needed and a method of building the parts into a successful antenna.

The length of the dipole is determined by the maximum space available. (frequency is also a factor but that discussion is beyond the scope of this text). The table below will be of great aid in building your dipole.

HALF WAVE DIPOLE ANTENNA MEASUREMENTS

FREQ. MHz.	AMATEUR BAND	TOTAL LENGTH	EACH SIDE
1		468 Ft 0 in	234 Ft 0 in
2		234 Ft 0 in	117 Ft 0 in
3		156 Ft 0 in	78 Ft 0 in
4	75/80	117 Ft 0 in	58 Ft 6 in
5		93 Ft 7 in	46 Ft 9 in
6		78 Ft 0 in	39 Ft 0 in
7	40	66 Ft 10 in	33 Ft 5 in
8		58 Ft 6 in	29 Ft 3 in
9		52 Ft 0 in	26 Ft 0 in
10	30	46 Ft 9 in	23 Ft 4 in
11		42 Ft 6 in	21 Ft 3 in
12		39 Ft 0 in	19 Ft 6 in
13		36 Ft 0 in	18 Ft 0 in
14	20	33 Ft 5 in	16 Ft 8 in
15		31 Ft 2 in	15 Ft 7 in
16		29 Ft 3 in	14 Ft 7 in
17		27 Ft 6 in	13 Ft 9 in
18		26 Ft 0 in	13 Ft 0 in
19		24 Ft 7 in	12 Ft 3 in
20		23 Ft 4 in	11 Ft 8 in
21	15	22 Ft 3 in	11 Ft 1 in
22		21 Ft 3 in	10 Ft 7 in
23		20 Ft 4 in	10 Ft 2 in
24	12	19 Ft 6 in	9 Ft 9 in
25		18 Ft 8 in	9 Ft 4 in
26		18 Ft 0 in	9 Ft 0 in
27		17 Ft 3 in	8 Ft 7 in
28	10	16 8 in	8 4 in
29	10	16 1 in	8 0 in
30		15 7 in	7 9 in

You will find that building the antenna in a "V" shape will improve the reception. Place the center insulation lower or higher that the sides insulators. Each side being 45 degrees off horizontal is the optimum angle.

Inverted "V" dipole antenna can also be used for transmitting on the C.B. band if you build it the proper size. (27 MHz = 8 ft 7 in. per side)

The Insulators can be any piece of non-conducting material like wood or plastic. The antenna wire can be almost any type of wire around your home. You may use bare or insulated wire. If you do use insulated wire, remember to strip back the insulation in the areas where you connect the coax wires.

CAUTION: If you are using bare wire (uninsulated), DO NOT touch the wire while transmitting. You may get electrocuted. Be sure that you don't position the antenna where someone else could come in contact with it.

You may use any 50 ohm coax. RG-58 is the most common. You may purchase the coax at Radio Shack or any other electronic store.

Another popular type of antenna is called the long wire. This antenna is basically a long piece of wire placed inside or outside your home. It operates on the assumption that the longer the wire, the more possibility of a section of the wire receiving the desired signal.

The long wire operates best when placed outside and extending in one direction as long as possible. The dipole antenna described above will, if constructed properly, perform much better that a long wire.

Another type of successful antenna is the full wave loop. Construction plans are located on the next page. It is built with the same parts as the dipole antenna.

This type of antenna will receive weak signals much better that almost any other type of antenna. The only drawback to the full-wave loop antenna is that it loses receiving ability outside its tuned size.

To explain this in a simpler manner, all antennas are cut or tuned to a specific frequency. It will operate best on signals around this frequency. It will receive all other frequencies in varying degrees. The simple rule of hand is, the longer the better.

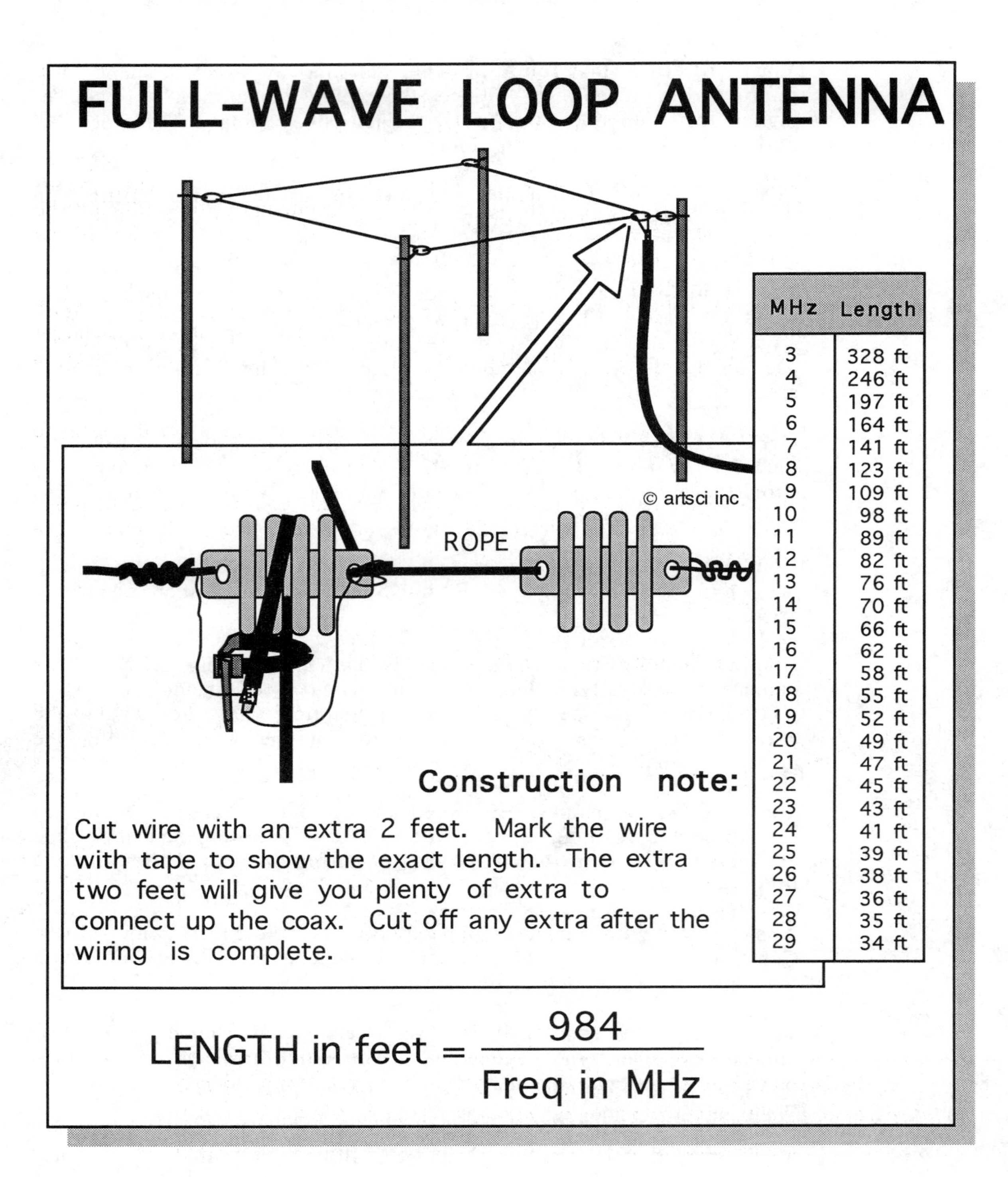

MHz	Length
3	328 ft
4	246 ft
5	197 ft
6	164 ft
7	141 ft
8	123 ft
9	109 ft
10	98 ft
11	89 ft
12	82 ft
13	76 ft
14	70 ft
15	66 ft
16	62 ft
17	58 ft
18	55 ft
19	52 ft
20	49 ft
21	47 ft
22	45 ft
23	43 ft
24	41 ft
25	39 ft
26	38 ft
27	36 ft
28	35 ft
29	34 ft

$$\text{LENGTH in feet} = \frac{984}{\text{Freq in MHz}}$$

Good luck and don't be afraid to experiment. There is nothing you can do with an antenna that will damage your radio. Unless of course you apply electricity or get hit by lighting. Remember, the bigger a hole you poke in the sky, the more signals you will capture.

Propagation

But I heard the station yesterday

Radio waves like light waves travel in straight lines. This means that if you can not "see" the transmitter antenna, you will not be able to hear the signal. Unless of course the signal is reflected by some object towards your direction.

Reflected signals are a great deal weaker than the original. With this fact in mind, the stronger the transmitted signal, the stronger the reflected signals will be. Also the more reflective the object is the better the strength of the reflected signal.

• Shortwave signals can be received around the world because they are reflected by the ocean and by the ionosphere.

The Ionosphere is a region of the sky between 60 and 250 miles above the earth's surface where ions and free electrons are generated by the sun's ultraviolet rays.

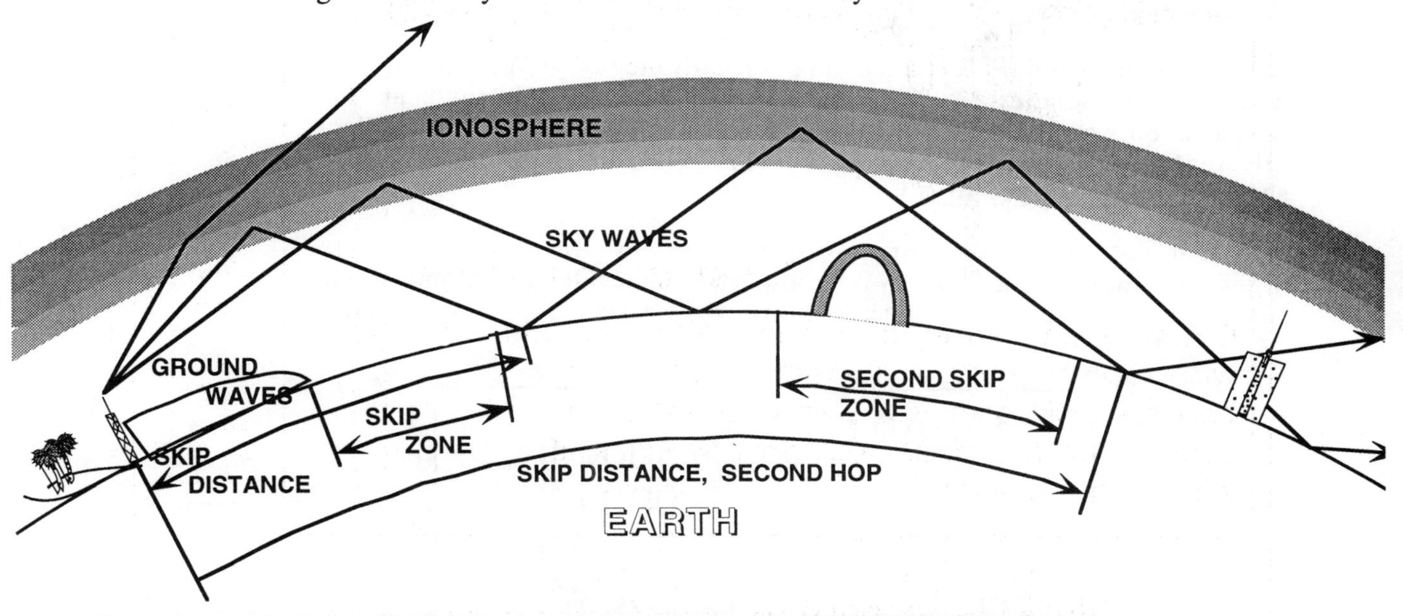

SKY WAVES BOUNCE OFF THE IONOSPHERE AND BACK TO EARTH. THIS ALLOWS THE WAVES TO TRAVEL GREAT DISTANCES.

Higher frequency (12-30 MHz) radio waves reflect or propagate better when the sun is shining and charging the ionosphere. The lower frequencies propagate better when the sun is not shining (it is on the other side of the earth).

Radio wave "skip" off the ionosphere in the same way a flat stone or rock can be skipped off the surface of a lake. Remember "skipping" rocks when you were a kid? The rock needs to hit the surface at a low angle to skip properly. Radio wave behave in the same way. The angle must be right to skip properly.

Sun spots, man-made noise and other conditions can effect the propagation ability of the ionosphere. Because of these conditions and other factors, the ionosphere may not always reflect the same signal in the same way from hour to hour or day to day. Using the rock skipping example again, the ionosphere is not a smooth lake surface, it is more like a wavy moving surface. There no telling how well the rock will skip.

As conditions change, the propagation varies and the directions of the signal will be changed. You may be able to hear a station for days in a row and then suddenly not be able to receive it for weeks of months at a time.

In the example of propagation on the previous page, the radio station can be heard by house "A" and house "C". House "B" can not hear the station because the signal skips over that section of the earth.

Many international broadcasters simultaneously broadcast on different frequencies. They do this in an attempt to beat the propagation problem. Since different frequencies will reflect differently at the same time of day, one of the transmissions should be able to get to all points on the earth.

You can test the quality of propagation at any time of the day or night by monitoring the WWV International Standards and Time frequencies. These transmissions are simulcasted 24 hours a day.

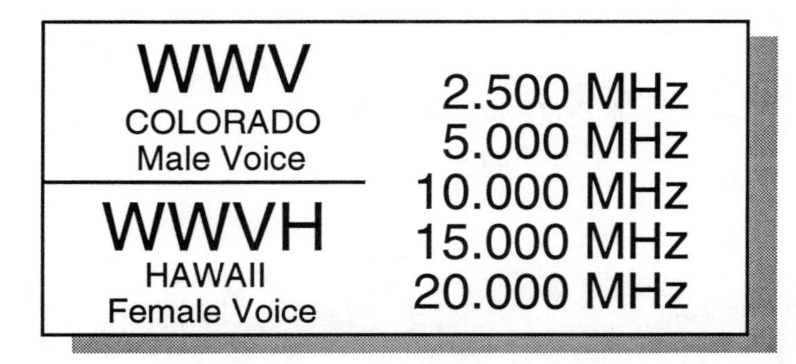

You'll be surprised as the quality of the signals change throughout the day.

Arctic Ocean
ICELAND
GREENLAND
Bering Sea
Beaufort Sea
Fairbanks
Anchorage
Baffin Bay
Godthab
Gulf of Alaska
Whitehorse
Radio Canada International
0515-0600 UTC 6.150, 9.050 MHz
Yellow Knife
Labrador Sea
Hudson Bay
North Pacific Ocean
CANADA
Churchill
Goose Bay
Edmonton
Vancouver
Christian Science Monitor
Calgary
North Atlantic Ocean
Seattle
Winnipeg
Quebec
Montreal
Halifax
Ottawa
0000-0200 UTC 9.850 MHz
0000-0455 UTC 13.760 MHz
0300-0600 UTC 7.395 MHz
0400-0800 UTC 9.840 MHz
0600-0855 UTC 11.705 MHz
0800-0855 UTC 13.760 MHz
1200-1255 UTC 9.475 MHz
1200-1400 UTC 9.425 MHz
1200-1455 UTC 13.760 MHz
1600-1655 UTC 11.580 MHz
1600-2400 UTC 17.555 MHz
1800-2000 UTC 21.545 MHz
2200-2255 UTC 9.465, 13.770 MHz
Toronto
Boston
Detroit
Salt Lake City
Chicago
New York
San Francisco
Denver
Pittsburgh
Washington
Voice of America
U. S. A.
Los Angeles
0000-0100 UTC 9.455 MHz
0000-0230 UTC 5.995, 11.580 MHz
0100-0300 UTC 7.115 MHz
1000-1200 UTC 5.985 MHz
1100-1400 UTC 15.155 MHz
1100-1500 UTC 6.110 MHz
2100-0100 UTC 17.735 MHz
Tijuana
Atlanta
El Paso
Houston
Hermosillo
Radio Mexico
Chihuahua
New Orleans
0200-0700 UTC 6.010 Mhz
0300-0500 UTC 15.230 MHz
Monterrey
Miami
Gulf of Mexico
MEXICO
La Paz
CUBA
Merida
Caribbean Sea
Guadalajara
Veracruz
BELIZE
DOM. REP.
Mexico City
HAITI
HONDURAS
Radio Cuba
Radio Costa Rica
Oaxaca
Acapulco
Tapachula
0000-0200 UTC 15.140 MHz
0000-0600 UTC 5.055 MHz
0100-0600 UTC 11.950 Mhz
GUATEMALA
0600-0800 UTC 11.760 MHz
Radio for Peace International
EL SALVADOR
2000-2200 UTC 13.700 MHz
2200 UTC 21.565 MHz
NICARAGUA
COST RICA
PANAMA

Austria

Radio Austria International
A-1136 Vienna, Austria

0130-0200 UTC 9.870 MHz
0530- 0600 UTC 6.015 MHz
0630-0700 UTC 6.015 MHz
1830-1900 UTC 5.945 MHz
2130-2200 UTC 5.945, 9.870 MHz

Belguim

Belgische Radio International
B-1000 Brussels, Belguim

Finland

Radio Finland
PO Box 10
SF-00241 Helsinki, Finland

0245-0345 UTC 9.560, 11.755 MHz
1230-1330 UTC11.735 MHz
1230-1430 UTC 15.400 MHz
1330-1530 UTC 21.550 MHz

France

Radio France Internationale
B.P. 9516 F-75016
Paris Cedex 16 , France

0200-0400 UTC 11.625 MHz
0300-0700 UTC 13.695 MHz
1300-1500 UTC 21.770 MHz
1600-2300 UTC 17.620 MHz
1800-2100 UTC 21.500 MHz

Germany

Deutsche Welle
P.O. Box 10 04 44
W-5000 Koln, Germany

0300-0500 UTC 6.145 , 9.545, 15.425 MHz
0400-0500 UTC 7.225 MHz
0500-0600 UTC 5.960, 13.610 MHz

Italy

RTV Italiana
Viale Mazzini 14
I-00195 Roma, Italy

0100-0120 UTC 11.800, 9.575 MHz
1935-1955 UTC 11.800, 9.710, 7.275

Netherlands

Radio Netherlands
Postbus 222 NL-1200 JG
Hilversum, Holland

0030-0125 UTC 6.020, 6.165 MHz
0330-0430 UTC 9.590 MHz
2000-2200 UTC 11.660 MHz

Norway

Radio Norway
0340
Oslo 3, Norway

0100-0300 UTC 9.615, 15.360
0300-0500 UTC 9.645 MHz
1300-1400 UTC 17.780 MHz
1500-1600 UTC 11.870 MHz
1600-1700 UTC 15.360 MHz
2400-0100 UTC 9.645 MHz

Portugal

Radio Portugal International
Rua S. Marcal 1
1200 Lisbia, Portugal

Spain

Radio Exterior De Espana
Apartado de Correo 156.202
E-28080 Madrid, Spain

0200-0500 UTC 9.530 MHz
0900-1900 UTC 6.125, 9.620, 11.945, 12.035, 15.380, 17.715, 17.845 MHz
1900-2300 UTC 15.110 MHz
2300-0500 UTC 6.055 MHz

Sweden

Radio Sweden
S-105 10 Stockholm, Sweden

1300-2030 UTC 17.740 MHz

Switzerland

Swiss Radio International
P.O. Box CH-3000
Bern 15 Switzerland

0000-0030 UTC 6.135, 9,650, 17.730 MHz
0200-0230 UTC 6.135, 9.650, 12.035 MHz
0400-0430 UTC 6.135, 9.885, 12.035 MHz

United Kingdom

See BBC Info Page

TIME CONVERSION CHART

U.T.C.	PST	PDST MST	MDST CST	CDST EST	EDST
0:00	4 pm	5 pm	6 pm	7 pm	8 pm
1:00	5 pm	6 pm	7 pm	8 pm	9 pm
2:00	6 pm	7 pm	8 pm	9 pm	10 pm
3:00	7 pm	8 pm	9 pm	10 pm	11 pm
4:00	8 pm	9 pm	10 pm	11 pm	Midnight
5:00	9 pm	10 pm	11 pm	Midnight	1 am
6:00	10 pm	11 pm	Midnight	1 am	2 am
7:00	11 pm	Midnight	1 am	2 am	3 am
8:00	Midnight	1 am	2 am	3 am	4 am
9:00	1 am	2 am	3 am	4 am	5 am
10:00	2 am	3 am	4 am	5 am	6 am
11:00	3 am	4 am	5 am	6 am	7 am
12:00	4 am	5 am	6 am	7 am	8 am
13:00	5 am	6 am	7 am	8 am	9 am
14:00	6 am	7 am	8 am	9 am	10 am
15:00	7 am	8 am	9 am	10 am	11 am
16:00	8 am	9 am	10 am	11 am	Noon
17:00	9 am	10 am	11 am	Noon	1 pm
18:00	10 am	11 am	Noon	1 pm	2 pm
19:00	11 am	Noon	1 pm	2 pm	3 pm
20:00	Noon	1 pm	2 pm	3 pm	4 pm
21:00	1 pm	2 pm	3 pm	4 pm	5 pm
22:00	2 pm	3 pm	4 pm	5 pm	6 pm
23:00	3 pm	4 pm	5 pm	6 pm	7 pm

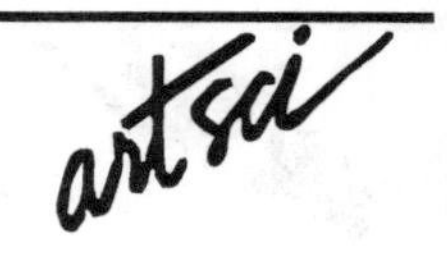

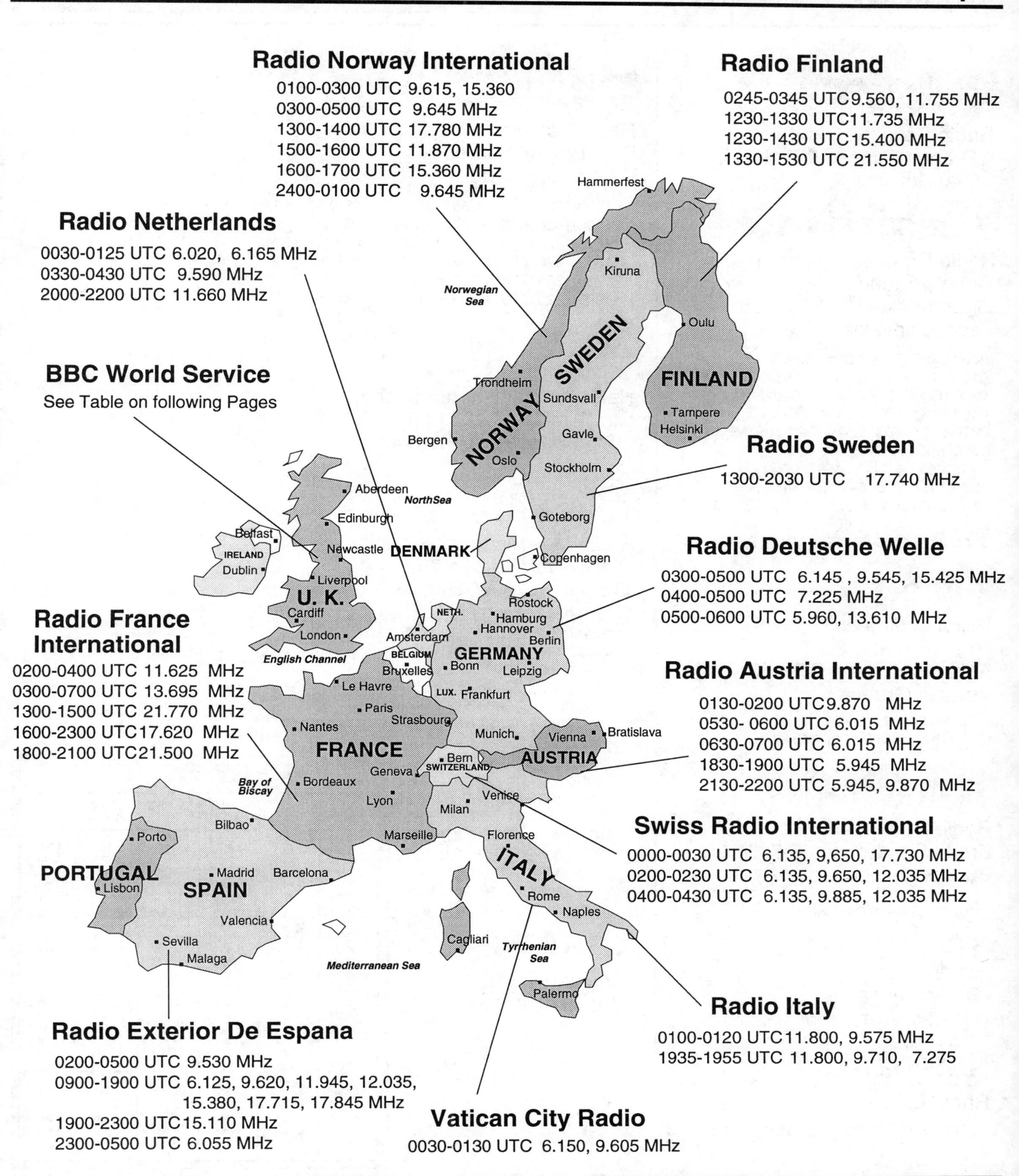
Radio Norway International
0100-0300 UTC 9.615, 15.360
0300-0500 UTC 9.645 MHz
1300-1400 UTC 17.780 MHz
1500-1600 UTC 11.870 MHz
1600-1700 UTC 15.360 MHz
2400-0100 UTC 9.645 MHz
Radio Finland
0245-0345 UTC 9.560, 11.755 MHz
1230-1330 UTC 11.735 MHz
1230-1430 UTC 15.400 MHz
1330-1530 UTC 21.550 MHz
Radio Netherlands
0030-0125 UTC 6.020, 6.165 MHz
0330-0430 UTC 9.590 MHz
2000-2200 UTC 11.660 MHz
BBC World Service
See Table on following Pages
Radio Sweden
1300-2030 UTC 17.740 MHz
Radio Deutsche Welle
0300-0500 UTC 6.145 , 9.545, 15.425 MHz
0400-0500 UTC 7.225 MHz
0500-0600 UTC 5.960, 13.610 MHz
Radio France International
0200-0400 UTC 11.625 MHz
0300-0700 UTC 13.695 MHz
1300-1500 UTC 21.770 MHz
1600-2300 UTC 17.620 MHz
1800-2100 UTC 21.500 MHz
Radio Austria International
0130-0200 UTC 9.870 MHz
0530- 0600 UTC 6.015 MHz
0630-0700 UTC 6.015 MHz
1830-1900 UTC 5.945 MHz
2130-2200 UTC 5.945, 9.870 MHz
Swiss Radio International
0000-0030 UTC 6.135, 9,650, 17.730 MHz
0200-0230 UTC 6.135, 9.650, 12.035 MHz
0400-0430 UTC 6.135, 9.885, 12.035 MHz
Radio Italy
0100-0120 UTC 11.800, 9.575 MHz
1935-1955 UTC 11.800, 9.710, 7.275
Radio Exterior De Espana
0200-0500 UTC 9.530 MHz
0900-1900 UTC 6.125, 9.620, 11.945, 12.035, 15.380, 17.715, 17.845 MHz
1900-2300 UTC 15.110 MHz
2300-0500 UTC 6.055 MHz
Vatican City Radio
0030-0130 UTC 6.150, 9.605 MHz
Hammerfest
Kiruna
Norwegian Sea
Oulu
SWEDEN
FINLAND
Trondheim
Sundsvall
Tampere
Helsinki
NORWAY
Bergen
Gavle
Oslo
Stockholm
Aberdeen
NorthSea
Edinburgh
Goteborg
Belfast
IRELAND
Dublin
Newcastle
DENMARK
Copenhagen
Liverpool
U. K.
Rostock
Cardiff
NETH.
Hamburg
Hannover
London
Amsterdam
Berlin
English Channel
BELGIUM
GERMANY
Bruxelles
Bonn
Leipzig
Le Havre
LUX.
Frankfurt
Paris
Strasbourg
Nantes
Munich
Vienna
Bratislava
FRANCE
Bern
AUSTRIA
SWITZERLAND
Geneva
Bay of Biscay
Bordeaux
Venice
Lyon
Milan
Bilbao
Marseille
Florence
Porto
ITALY
PORTUGAL
Madrid
Barcelona
Lisbon
SPAIN
Rome
Naples
Valencia
Cagliari
Sevilla
Tyrrhenian Sea
Malaga
Mediterranean Sea
Palermo

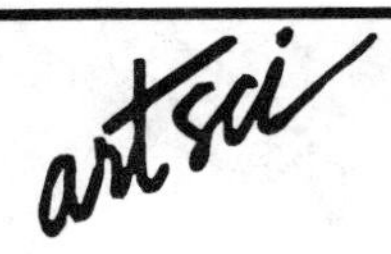
artsci

Bulgaria

Radio Sofia
4 Dragan Tsankov Blvd
Sofia, Bulgaria

Czechoslovakia

Radio Prague International
12099 Prague 2
Vinohradska 12
Czechoslovakia

0000-0027 UTC 7.345, 9.540, 11.990 MHz
0100-0130 UTC 5.930, 7.345, 9.540 MHz
0300-0330 UTC 7.345, 9.540 MHz
0400-0430 UTC 7.345, 9.540 MHz
0730-0800 UTC 17.725, 21.705 MHz
1800-1827 UTC 5.930, 6.055, 7.345, 9.605 MHz
1930-1957 UTC 6.055, 7.345 MHz
2100-2130 UTC 5.930, 6.055, 7.345, 9.605 MHz
2200-2225 UTC 6.055, 7.345, 9.605 MHz

Greece

Voice of Greece
Mesogion 432 Str.
Aghia Paraskevi GR-153 42
Athens, Greece

0130-0200 UTC 11.645 MHz
0430 UTC 11.940 MHz

Hungary

Radio Budapest
Brody Sandor utca 5-7 H-1800
Budapest, Hungary

0100 UTC 9.835 MHz

Latvia

Radio Latvia

0300-0500 UTC 5.935 MHz

Lithuania

Radio Lithuania

0600-0800 UTC 9.710 MHz

Poland

Radio Polonia
P.O. Box 46, 00-950
Warsaw, Poland

1300-1355 UTC 1.503, 6.305, 7.145, 9.525, 11.815 MHz
1400-1425 UTC 1.503, 6.135, 7.285 MHz
1500-1525 UTC 1.503, 6.095, 7.145, 9.540 MHz
1600-1625 UTC 1.503, 9.540, 6.135 MHz
1730-1755 UTC 1.503, 6.095, 6.135, 7.285, 9.525 MHz
1900-1925 UTC 1.503, 6.135, 7.145 MHz

Romania

Radio Romania International
Str. G-ral Berthelot nr. 60-62
79756 Bucharest, Romania

Russia

Radio Moscow
ulitsia Pyatnitskaya 25
113326 Moscow, Russia

0000-0300 UTC 12.050, 15.290 MHz
0000-0500 UTC 11.850, 12.040 MHz
0000-0600 UTC 17.665 MHz
0330-0500 UTC 12.010 MHz
0400-0500 UTC 11.980 MHz
0400-0600 UTC 17.655 MHz
1000-2200 UTC 11.840 MHz
1300-0000 UTC 12.050 MHz
1300-1800 UTC 11.995 MHz
1300-1900 UTC 17.670 MHz
1500-1700 UTC 12.030 MHz
1500-2200 UTC 13.645 MHz
1600-2200 UTC 15.375 MHz
1900-2200 UTC 12.070 MHz
2000-0000 UTC 15.355 MHz
2100-2200 UTC 12.040 MHz

Ukraine

Ukrainian Radio
Address unknown

TIME CONVERSION CHART

U.T.C.	PST	PDST MST	MDST CST	CDST EST	EDST
0:00	4 pm	5 pm	6 pm	7 pm	8 pm
1:00	5 pm	6 pm	7 pm	8 pm	9 pm
2:00	6 pm	7 pm	8 pm	9 pm	10 pm
3:00	7 pm	8 pm	9 pm	10 pm	11 pm
4:00	8 pm	9 pm	10 pm	11 pm	Midnight
5:00	9 pm	10 pm	11 pm	Midnight	1 am
6:00	10 pm	11 pm	Midnight	1 am	2 am
7:00	11 pm	Midnight	1 am	2 am	3 am
8:00	Midnight	1 am	2 am	3 am	4 am
9:00	1 am	2 am	3 am	4 am	5 am
10:00	2 am	3 am	4 am	5 am	6 am
11:00	3 am	4 am	5 am	6 am	7 am
12:00	4 am	5 am	6 am	7 am	8 am
13:00	5 am	6 am	7 am	8 am	9 am
14:00	6 am	7 am	8 am	9 am	10 am
15:00	7 am	8 am	9 am	10 am	11 am
16:00	8 am	9 am	10 am	11 am	Noon
17:00	9 am	10 am	11 am	Noon	1 pm
18:00	10 am	11 am	Noon	1 pm	2 pm
19:00	11 am	Noon	1 pm	2 pm	3 pm
20:00	Noon	1 pm	2 pm	3 pm	4 pm
21:00	1 pm	2 pm	3 pm	4 pm	5 pm
22:00	2 pm	3 pm	4 pm	5 pm	6 pm
23:00	3 pm	4 pm	5 pm	6 pm	7 pm

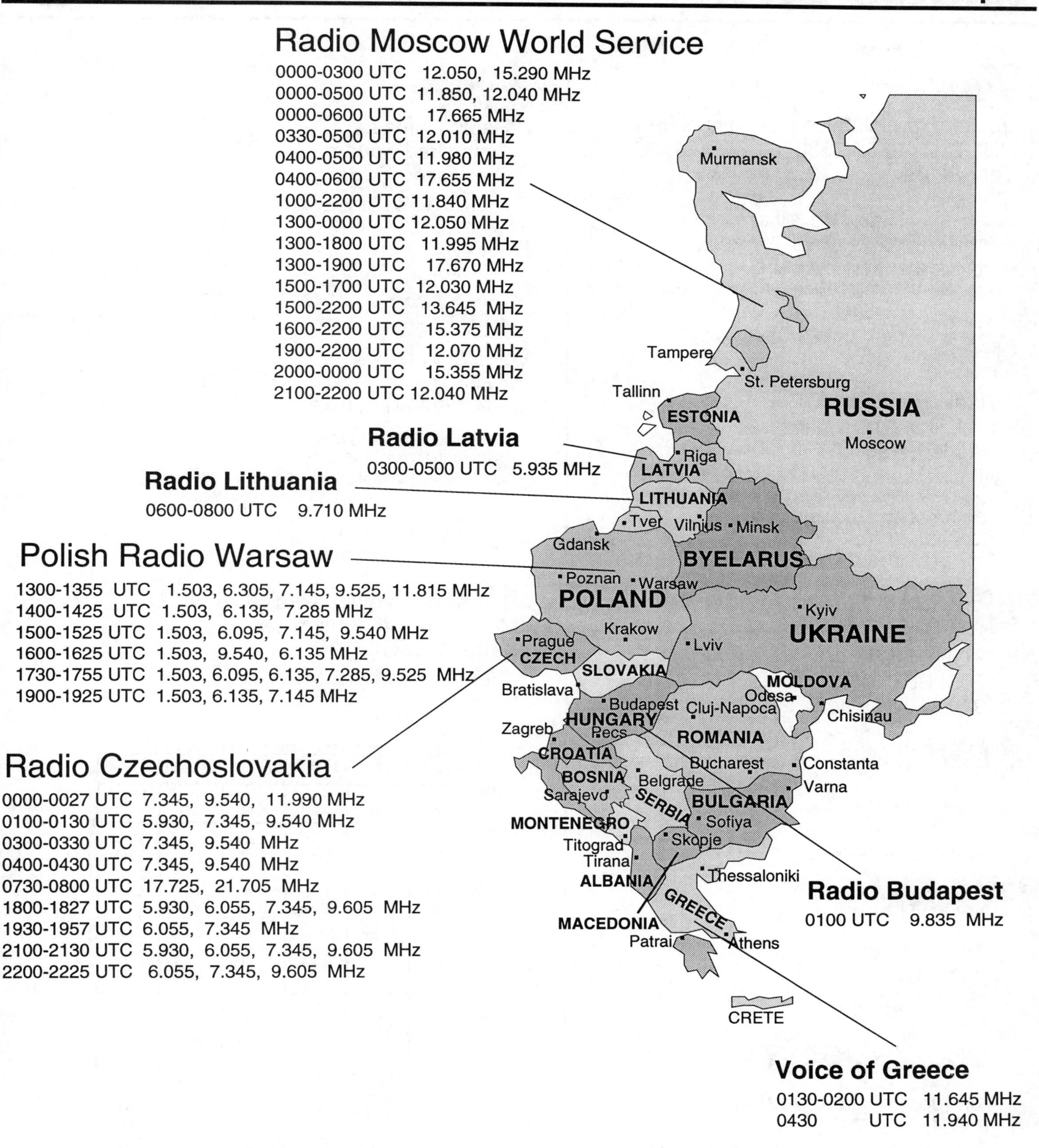
Radio Moscow World Service
0000-0300 UTC 12.050, 15.290 MHz
0000-0500 UTC 11.850, 12.040 MHz
0000-0600 UTC 17.665 MHz
0330-0500 UTC 12.010 MHz
0400-0500 UTC 11.980 MHz
0400-0600 UTC 17.655 MHz
1000-2200 UTC 11.840 MHz
1300-0000 UTC 12.050 MHz
1300-1800 UTC 11.995 MHz
1300-1900 UTC 17.670 MHz
1500-1700 UTC 12.030 MHz
1500-2200 UTC 13.645 MHz
1600-2200 UTC 15.375 MHz
1900-2200 UTC 12.070 MHz
2000-0000 UTC 15.355 MHz
2100-2200 UTC 12.040 MHz
Radio Latvia
0300-0500 UTC 5.935 MHz
Radio Lithuania
0600-0800 UTC 9.710 MHz
Polish Radio Warsaw
1300-1355 UTC 1.503, 6.305, 7.145, 9.525, 11.815 MHz
1400-1425 UTC 1.503, 6.135, 7.285 MHz
1500-1525 UTC 1.503, 6.095, 7.145, 9.540 MHz
1600-1625 UTC 1.503, 9.540, 6.135 MHz
1730-1755 UTC 1.503, 6.095, 6.135, 7.285, 9.525 MHz
1900-1925 UTC 1.503, 6.135, 7.145 MHz
Radio Czechoslovakia
0000-0027 UTC 7.345, 9.540, 11.990 MHz
0100-0130 UTC 5.930, 7.345, 9.540 MHz
0300-0330 UTC 7.345, 9.540 MHz
0400-0430 UTC 7.345, 9.540 MHz
0730-0800 UTC 17.725, 21.705 MHz
1800-1827 UTC 5.930, 6.055, 7.345, 9.605 MHz
1930-1957 UTC 6.055, 7.345 MHz
2100-2130 UTC 5.930, 6.055, 7.345, 9.605 MHz
2200-2225 UTC 6.055, 7.345, 9.605 MHz
Radio Budapest
0100 UTC 9.835 MHz
Voice of Greece
0130-0200 UTC 11.645 MHz
0430 UTC 11.940 MHz
Murmansk
Tampere
St. Petersburg
Tallinn
ESTONIA
RUSSIA
Moscow
Riga
LATVIA
LITHUANIA
Tver
Vilnius
Minsk
Gdansk
BYELARUS
Poznan
Warsaw
POLAND
Kyiv
UKRAINE
Krakow
Lviv
Prague
CZECH
SLOVAKIA
MOLDOVA
Bratislava
Odesa
Budapest
Cluj-Napoca
Chisinau
HUNGARY
Zagreb
Pecs
ROMANIA
CROATIA
Bucharest
Constanta
BOSNIA
Belgrade
Varna
Sarajevo
SERBIA
BULGARIA
MONTENEGRO
Sofiya
Titograd
Skopje
Tirana
Thessaloniki
ALBANIA
GREECE
MACEDONIA
Patrai
Athens
CRETE

artsci

Iran

Voice of the Islamicc Republic of Iran
P.O. Box 19395/3333
Tehran, Iran

1130-1230 UTC 1.224 , 7.215, 9.575 , 9.695 , 11.790, 11.930 MHz
1930-2030 UTC 9.022 MHz

1400-1500 UTC 6.035, 7.220 MHz
1130-1225 UTC 9.705 MHz

Iraq

Radio Bagdad
P.O. Box 8145
Bagdad, Iraq

0230 UTC 9.400 MHz
0400 UTC 7.475 , 13.655 MHz
0430-0730 UTC 7.095 MHz
1730-2030 UTC 7.095 MHz

Israel

KOL Israel
P.O. Box 1082
91010 Jerusalem, Israel

Lebanon

Voice of Hope
P.O. Box 3379
Limassol, Cyprus

Oman

Radio Oman
P.O. Box 600
Muscat, Oman

Saudi Arabia

Kingdom of Saudi Arabia
P.O. Box 61718
Riyadh 11575, Saudia Arabia

9.705, 9.720 MHz
0400 UTC 15.060 MHz

Syria

Radio Damascus
Ommayad Square
Damascus, Syria

2100 UTC 9.400 MHz

Turkey

Voice of Turkey
P.K. 333
06-443 Ankara, Turkey

0355-0500 UTC 9.445 MHz
1330-1400 UTC 17.785 MHz
2300-2400 UTC 9.445, 9.665, 9.685, 17.760 MHz

Yemen

Republic of Yemen Radio
Ministry of information
San'a, Yemen

0800-1400 UTC 9.780 MHz

TIME CONVERSION CHART

U.T.C.	PST	PDST MST	MDST CST	CDST EST	EDST
0:00	4 pm	5 pm	6 pm	7 pm	8 pm
1:00	5 pm	6 pm	7 pm	8 pm	9 pm
2:00	6 pm	7 pm	8 pm	9 pm	10 pm
3:00	7 pm	8 pm	9 pm	10 pm	11 pm
4:00	8 pm	9 pm	10 pm	11 pm	Midnight
5:00	9 pm	10 pm	11 pm	Midnight	1 am
6:00	10 pm	11 pm	Midnight	1 am	2 am
7:00	11 pm	Midnight	1 am	2 am	3 am
8:00	Midnight	1 am	2 am	3 am	4 am
9:00	1 am	2 am	3 am	4 am	5 am
10:00	2 am	3 am	4 am	5 am	6 am
11:00	3 am	4 am	5 am	6 am	7 am
12:00	4 am	5 am	6 am	7 am	8 am
13:00	5 am	6 am	7 am	8 am	9 am
14:00	6 am	7 am	8 am	9 am	10 am
15:00	7 am	8 am	9 am	10 am	11 am
16:00	8 am	9 am	10 am	11 am	Noon
17:00	9 am	10 am	11 am	Noon	1 pm
18:00	10 am	11 am	Noon	1 pm	2 pm
19:00	11 am	Noon	1 pm	2 pm	3 pm
20:00	Noon	1 pm	2 pm	3 pm	4 pm
21:00	1 pm	2 pm	3 pm	4 pm	5 pm
22:00	2 pm	3 pm	4 pm	5 pm	6 pm
23:00	3 pm	4 pm	5 pm	6 pm	7 pm

The Voice of Turkey

0355-0500 UTC 9.445 MHz
1330-1400 UTC 17.785 MHz
2300-2400 UTC 9.445, 9.665, 9.685, 17.760 MHz

Islamic Republic of Iran

1130-1230 UTC 1.224 , 7.215, 9.575 , 9.695 , 11.790, 11.930 MHz
1930-2030 UTC 9.022 MHz

Voice of Irag

0230 UTC 9.400 MHz
0400 UTC 7.475 , 13.655 MHz
0430-0730 UTC 7.095 MHz
1730-2030 UTC 7.095 MHz

The Voice of Iran

1400-1500 UTC 6.035, 7.220 MHz
1130-1225 UTC 9.705 MHz

Radio Damascus

2100 UTC 9.400 MHz

The Kingdom of Saudi Arabia

9.705, 9.720 MHz
0400 UTC 15.060 MHz

Radio Yemen

0800-1400 UTC 9.780 MHz

Australia

Radio Australia
P.O. Box 755
Glen Waverley VIC 3150 Australia

0000-0900 UTC 15.240, 21.740 MHz
0200-0730 UTC 21.525 MHz
0300-0600 UTC 17.715 MHz
0500-0600 UTC 15.170 MHz
0800-1100 UTC 15.160 MHz
0900-1000 UTC 7.140, 13.605, 15.170 MHz
0900-1200 UTC 17.715 MHz
1100-1200 UTC 7.140, 13.605, 15.170 MHz
1100-1430 UTC 21.720 MHz
1100-2100 UTC 6.080 MHz
1130-1530 UTC 11.720 MHz
1130-2100 UTC 5.995 MHz
1430-2100 UTC 12.000, 13.755 MHz
1600-2030 UTC 6.060, 11.910 MHz
1600-2330 UTC 13.605 MHz
2000-0800 UTC 11.930 MHz
2000-2400 UTC 17.995 MHz
2100-0700 UTC 15.160 MHz
2100-0800 UTC 11.880 MHz
2100-2300 UTC 13.705 MHz
2130-2400 UTC 17.715 MHz
2200-0400 UTC 21.740 MHz
2200-0700 UTC 15.365 MHz

China

Radio Beijing
No 2. Fuxingmenwai
Beijing, China

0000-0100 UTC 9.770, 11.715 MHz
0300-0400 UTC 9.770, 11.715 MHz
0900-1000 UTC 17.710 MHz
0900-1100 UTC 15.440 MHz
1200-1300 UTC 9.665 MHz
1200-1400 UTC 11.600 MHz
1400-1600 UTC 7.405 MHz

India

Radio India
P.O. Box 500
New Delhi 110 001 India

0630 UTC 9.675 MHz
2000-2200 UTC 11.620 MHz

Japan

Radio Japan
2-2-1 Jinnan Shibuya-ku
Tokyo, Japan

0300-0330 UTC 15.325, 21.610 MHz
1200-1230 UTC 15.325, 21.610 MHz

Philippines

Far East Broadcasting
P.O. Box 1
Valenzuela, Metro Manila

South Korea

Radio Korea
46 Yo-ui-do-dong
Yongdungo'o-gu
Seoul 150-790 Republic of Korea

0600-0700 UTC 7.275 11.810 MHz
0800-0900 UTC 7.550, 13.670 MHz
1100-1200 UTC 15.575 MHz
1130-1200 UTC 9.650 MHz
1215-1315 UTC 9.750 , 9.570 MHz
1400-1500 UTC 9.570 MHz
1600-1700 UTC 9.820 MHz
1800-1900 UTC 15.575 MHz
1930-2000 UTC 1.170, 6.135 MHz
2030-2130 UTC 6.480, 7.550, 15.575 MHz
2400-0100 UTC 15.575 MHz

Thailand

Radio Thailand
Bankkok 10200 Thailand

Vietnam

Voice of Vietnam
Overseas Service
58 Quan Su
Hanoi, Vietnam

1100-1130 UTC 7.416, 9.732 MHz
1000-1030 UTC 9.840, 12.020, 15.010 MHz
1230-1300 UTC 9.840, 12.020, 15.010 MHz
1330-1400 UTC 9.840, 12.020, 15.010 MHz
1600-1630 UTC 9.840, 12.020, 15.010 MHz
1800-1830 UTC 9.840, 12.020, 15.010 MHz
1900-1930 UTC 9.840, 12.020, 15.010 MHz
2030-2100 UTC 9.840, 12.020, 15.010 MHz
2330-2400 UTC 9.840, 12.020, 15.010 MHz

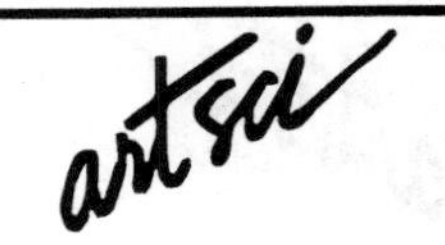

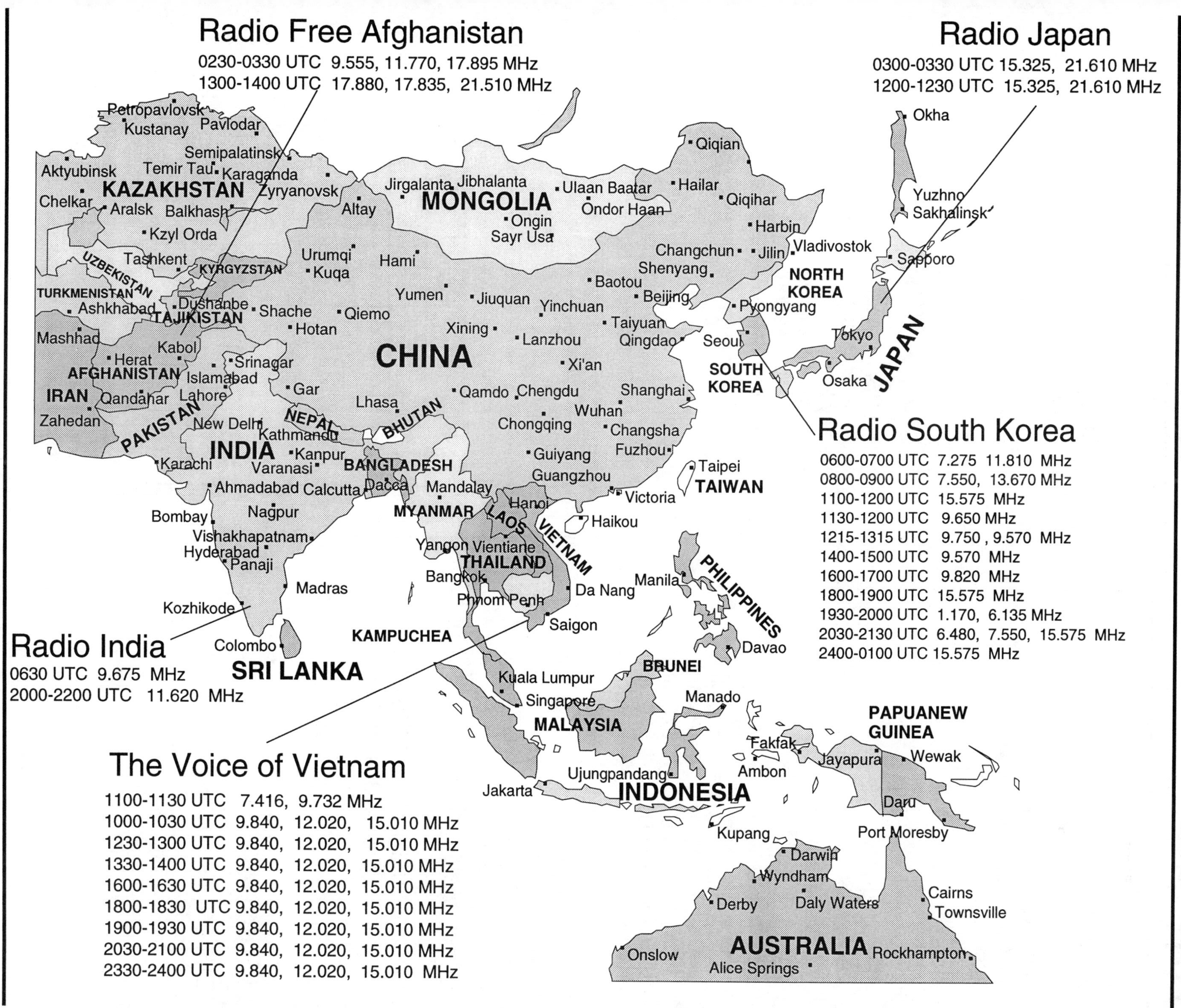
Radio Free Afghanistan
0230-0330 UTC 9.555, 11.770, 17.895 MHz
1300-1400 UTC 17.880, 17.835, 21.510 MHz
Radio Japan
0300-0330 UTC 15.325, 21.610 MHz
1200-1230 UTC 15.325, 21.610 MHz
Radio South Korea
0600-0700 UTC 7.275 11.810 MHz
0800-0900 UTC 7.550, 13.670 MHz
1100-1200 UTC 15.575 MHz
1130-1200 UTC 9.650 MHz
1215-1315 UTC 9.750 , 9.570 MHz
1400-1500 UTC 9.570 MHz
1600-1700 UTC 9.820 MHz
1800-1900 UTC 15.575 MHz
1930-2000 UTC 1.170, 6.135 MHz
2030-2130 UTC 6.480, 7.550, 15.575 MHz
2400-0100 UTC 15.575 MHz
Radio India
0630 UTC 9.675 MHz
2000-2200 UTC 11.620 MHz
The Voice of Vietnam
1100-1130 UTC 7.416, 9.732 MHz
1000-1030 UTC 9.840, 12.020, 15.010 MHz
1230-1300 UTC 9.840, 12.020, 15.010 MHz
1330-1400 UTC 9.840, 12.020, 15.010 MHz
1600-1630 UTC 9.840, 12.020, 15.010 MHz
1800-1830 UTC 9.840, 12.020, 15.010 MHz
1900-1930 UTC 9.840, 12.020, 15.010 MHz
2030-2100 UTC 9.840, 12.020, 15.010 MHz
2330-2400 UTC 9.840, 12.020, 15.010 MHz
Petropavlovsk
Kustanay
Pavlodar
Semipalatinsk
Aktyubinsk
Temir Tau
Karaganda
KAZAKHSTAN
Zyryanovsk
Chelkar
Aralsk
Balkhash
Kzyl Orda
Tashkent
UZBEKISTAN
KYRGYZSTAN
TURKMENISTAN
Ashkhabad
Dushanbe
TAJIKISTAN
Mashhad
Kabol
Herat
AFGHANISTAN
Islamabad
Srinagar
IRAN
Qandahar
Lahore
Zahedan
PAKISTAN
New Delhi
NEPAL
Kathmandu
Karachi
INDIA
Kanpur
Varanasi
BANGLADESH
Ahmadabad
Calcutta
Dacca
Bombay
Nagpur
Vishakhapatnam
Hyderabad
Panaji
Madras
Kozhikode
Colombo
SRI LANKA
Jirgalanta
Jibhalanta
MONGOLIA
Ulaan Baatar
Ondor Haan
Ongin
Sayr Usa
Altay
Urumqi
Kuqa
Hami
Shache
Qiemo
Hotan
Yumen
Jiuquan
Yinchuan
Xining
Lanzhou
CHINA
Gar
Lhasa
BHUTAN
Qamdo
Chengdu
Chongqing
Wuhan
Xi'an
Baotou
Beijing
Taiyuan
Qingdao
Shanghai
Changsha
Fuzhou
Guiyang
Guangzhou
Victoria
Haikou
Qiqian
Hailar
Qiqihar
Harbin
Changchun
Jilin
Vladivostok
Shenyang
NORTH KOREA
Pyongyang
Seoul
SOUTH KOREA
Okha
Yuzhno Sakhalinsk
Sapporo
Tokyo
Osaka
JAPAN
Taipei
TAIWAN
Mandalay
MYANMAR
Hanoi
LAOS
VIETNAM
Yangon
Vientiane
THAILAND
Bangkok
Phnom Penh
Da Nang
Saigon
KAMPUCHEA
Manila
PHILIPPINES
Davao
Kuala Lumpur
Singapore
BRUNEI
MALAYSIA
Manado
Fakfak
Ambon
Ujungpandang
Jakarta
INDONESIA
Kupang
PAPUANEW GUINEA
Jayapura
Wewak
Daru
Port Moresby
Darwin
Wyndham
Derby
Daly Waters
Cairns
Townsville
AUSTRALIA
Onslow
Alice Springs
Rockhampton

Algeria

RTV Algerienne
21 Boulevard des Martyrs
Algiers, Algeria

Egypt

Radio Cairo
P.O. Box 1186
Cairo, Egypt

0200-0330 UTC 9.475 MHz
0200-0300 UTC 9.675 MHz
1650-1705 UTC 11.975 MHz
2100-2300 UTC 9.900 MHz

Ethiopia

Voice of Ethiopia
P.O. Box 654
Addia Ababa, Ethiopia

1500-1550 UTC 9.560 MHz
0400-0500 UTC 9.705 MHz

Guinea

RTV Guineenne
B.P. 391
Conakry, Guinea

2000-2300 UTC 7.190 MHz

Morocco

Radio Morocco

1400-1600 UTC 17.595 MHz

South Africa

Radio RSA
Piet Meyer Building
Henley Raod
Broadcasting Centre
Johannesburg 2000
Republic of South Africa

Zaire

La Voix du Zaire
Kinshasa
B.P. 3171
Kinshasa-Gombe, Zaire

TIME CONVERSION CHART

U.T.C.	PST	PDST MST	MDST CST	CDST EST	EDST
0:00	4 pm	5 pm	6 pm	7 pm	8 pm
1:00	5 pm	6 pm	7 pm	8 pm	9 pm
2:00	6 pm	7 pm	8 pm	9 pm	10 pm
3:00	7 pm	8 pm	9 pm	10 pm	11 pm
4:00	8 pm	9 pm	10 pm	11 pm	Midnight
5:00	9 pm	10 pm	11 pm	Midnight	1 am
6:00	10 pm	11 pm	Midnight	1 am	2 am
7:00	11 pm	Midnight	1 am	2 am	3 am
8:00	Midnight	1 am	2 am	3 am	4 am
9:00	1 am	2 am	3 am	4 am	5 am
10:00	2 am	3 am	4 am	5 am	6 am
11:00	3 am	4 am	5 am	6 am	7 am
12:00	4 am	5 am	6 am	7 am	8 am
13:00	5 am	6 am	7 am	8 am	9 am
14:00	6 am	7 am	8 am	9 am	10 am
15:00	7 am	8 am	9 am	10 am	11 am
16:00	8 am	9 am	10 am	11 am	Noon
17:00	9 am	10 am	11 am	Noon	1 pm
18:00	10 am	11 am	Noon	1 pm	2 pm
19:00	11 am	Noon	1 pm	2 pm	3 pm
20:00	Noon	1 pm	2 pm	3 pm	4 pm
21:00	1 pm	2 pm	3 pm	4 pm	5 pm
22:00	2 pm	3 pm	4 pm	5 pm	6 pm
23:00	3 pm	4 pm	5 pm	6 pm	7 pm

Radio Cairo
0200-0330 UTC 9.475 MHz
0200-0300 UTC 9.675 MHz
1650-1705 UTC 11.975 MHz
2100-2300 UTC 9.900 MHz
Radio Morocco
1400-1600 UTC 17.595 MHz
Radio Ethiopia
0400-0500 UTC 9.705 MHz
Voice of Ethiopia
1500-1550 UTC 9.560 MHz
Radio Africa
2000-2300 UTC 7.190 MHz
Mediterranean Sea
Indian Ocean
South Atlantic
MOROCCO
ALGERIA
LIBYA
EGYPT
WESTERN SAHARA
MAURITANIA
MALI
NIGER
CHAD
SUDAN
ERITREA
DJIBOUTI
ETHIOPIA
SOMALIA
SENEGAL
GAMBIA
GUINEA BISSAU
GUINEA
SIERRA LEONE
LIBERIA
IVORY COAST
BURKINA
GHANA
TOGO
BENIN
NIGERIA
CAMEROON
CENTRAL AFRICAN REPUBLIC
EQUATORIAL GUINEA
SAO TOME & PRINCIPE
GABON
CONGO
ZAIRE
UGANDA
KENYA
RWANDA
BURUNDI
TANZANIA
Cabinda(ANGOLA)
ANGOLA
ZAMBIA
MALAWI
MOZAMBIQUE
ZIMBABWE
NAMIBIA
BOTSWANA
SOUTH AFRICA(Walvis Bay)
SOUTH AFRICA
SWAZILAND
LESOTHO
MADAGASCAR
TUNISIA
Tangier
Rabat
Casablanca
Marrakech
Constantine
Algiers
Oran
Ghardaia
Tunis
Tripoli
Banghazi
Alexandria
Cairo
La'youn
Tindouf
Al Jawf
Aswan
Tamanrasset
Nouakchott
Dakar
Banjul
Bissau
Conakry
Freetown
Monrovia
Abidjan
Nema
Tombouctou
Bamako
Ouagadougou
Niamey
Agadez
Zinder
Faya-Largeau
Lake Chad
N'Djamena
Kano
Maiduguri
Lome
Porto Novo
Lagos
Accra
Port Sudan
Khartoum
El Fashir
Asmera
Djibouti
Berbera
Addis Ababa
Wau
Juba
Lake Turkana
Douala
Malabo
Yaounde
Bangui
Bata
Sao Tome
Libreville
Lake Albert
Kampala
Kisangani
Lake Victoria
Nairobi
Mogadishu
Kigali
Bujumbura
Brazzaville
Kinshasa
Pointe-Noire
Kananga
Kalemie
Lake Tanganyika
Mombasa
Dar es Salaam
Luanda
Mbeya
Lubumbashi
Kasama
Lobito
Luena
Lake Nyasa
Kitwe
Lilongwe
Lusaka
Lake Kariba
Nacala
Namibe
Harare
Mahajanga
Antananarivo
Bulawayo
Beira
Mauritius
Port Lewis
Saint Denis
Reunion(FRANCE)
Windhoek
Gaborone
Toliara
Pretoria
Johannesburg
Maputo
Mbabane
Luderitz
Maseru
Durban
Port Elizabeth
Cape Town

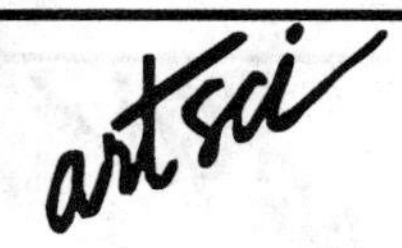
artsci

Argentina

Radio Argentinia
Casilla de Correo 2868
1000 Buenos Aires, Argentina

0100-0400 UTC 11.710 MHz

Bolivia

Radio Santa Cruz
Emisora del Instituto Radiofonoco
Casilla 672
Santa Cruz de la Sierra, Bolivia

1000-1200 UTC 6.135 MHz

Brazil

Radio Brazil
C.P. 405
78001 Cuiaba
Mato Grosso, Brazil

Chile

Radio Chile

0200-0300 UTC 11.810 MHz
2000-2200 UTC 15.140 MHz

Colombia

Radio Nacional de Colombia
AV. El Dorado
Bogota, Columbia

Ecuador

Radio Ecuador

1600-2300 UTC 21.480 MHz
1900-2200 UTC 17.790 MHz
0600-0700 UTC 15.115 MHz
0100-1100 UTC 9.745 MHz
0000-0500 UTC 5.040 MHz

French Guiana

RFO-Guyane
Cayenne
Frence Guyana

0330 UTC 5.325 MHz

Guyana

Voice of Guyana
P.O. Box 10760
Georgetown, Guyana

0730-0200 UTC. 5.950 MHz

Paraguay

Radio Nacional
Calle Montevideo
esq. Estrella
Asuncion, Paraguay

Peru

Radio Peru

0200-0300 UTC 4.790 MHz
0600-1100 UTC 5.095 MHz

Suriname

Radio Suriname International
Postbus 2979
Paramararibo, Suriname

Venezuela

Radio Venezuela

0300-0500 UTC 4.980 MHz
0300-0600 UTC 4.970 MHz

TIME CONVERSION CHART

U.T.C.	PST	PDST MST	MDST CST	CDST EST	EDST
0:00	4 pm	5 pm	6 pm	7 pm	8 pm
1:00	5 pm	6 pm	7 pm	8 pm	9 pm
2:00	6 pm	7 pm	8 pm	9 pm	10 pm
3:00	7 pm	8 pm	9 pm	10 pm	11 pm
4:00	8 pm	9 pm	10 pm	11 pm	Midnight
5:00	9 pm	10 pm	11 pm	Midnight	1 am
6:00	10 pm	11 pm	Midnight	1 am	2 am
7:00	11 pm	Midnight	1 am	2 am	3 am
8:00	Midnight	1 am	2 am	3 am	4 am
9:00	1 am	2 am	3 am	4 am	5 am
10:00	2 am	3 am	4 am	5 am	6 am
11:00	3 am	4 am	5 am	6 am	7 am
12:00	4 am	5 am	6 am	7 am	8 am
13:00	5 am	6 am	7 am	8 am	9 am
14:00	6 am	7 am	8 am	9 am	10 am
15:00	7 am	8 am	9 am	10 am	11 am
16:00	8 am	9 am	10 am	11 am	Noon
17:00	9 am	10 am	11 am	Noon	1 pm
18:00	10 am	11 am	Noon	1 pm	2 pm
19:00	11 am	Noon	1 pm	2 pm	3 pm
20:00	Noon	1 pm	2 pm	3 pm	4 pm
21:00	1 pm	2 pm	3 pm	4 pm	5 pm
22:00	2 pm	3 pm	4 pm	5 pm	6 pm
23:00	3 pm	4 pm	5 pm	6 pm	7 pm

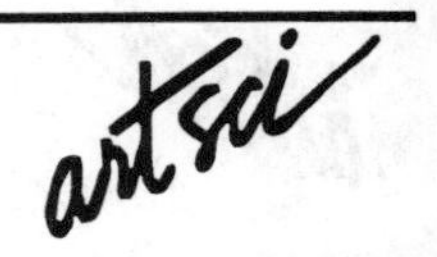

Radio Ecuador
1600-2300 UTC 21.480 MHz
1900-2200 UTC 17.790 MHz
0600-0700 UTC 15.115 MHz
0100-1100 UTC 9.745 MHz
0000-0500 UTC 5.040 MHz
Radio Rumbos
0300-0500 UTC 4.980 MHz
Radio Venezuela
0300-0600 UTC 4.970 MHz
Voice of Guyana
0730-0200 UTC. 5.950 MHz
French Guiana
0330 UTC 5.325 MHz
Radio Peru
0200-0300 UTC 4.790 MHz
0600-1100 UTC 5.095 MHz
Radio Santa Cruz
1000-1200 UTC 6.135 MHz
Weather
10.225 MHz
Radio Chile
0200-0300 UTC 11.810 MHz
2000-2200 UTC 15.140 MHz
Radio Argentina
0100-0400 UTC 11.710 MHz
VENEZUELA
GUYANA
SURINAME
FRENCH GUIANA (FRANCE)
COLOMBIA
ECUADOR
PERU
BOLIVIA
BRAZIL
PARAGUAY
CHILE
URUGUAY
ARGENTINA
FALKLAND ISLANDS (U.K.)
SOUTH GEORGIA ISLAND (U.K.)
Barranquilla
Caracas
Ciudad Guayana
Cucuta
San Cristobal
Georgetown
Paramaribo
Cayenne
Medellin
Bogota
Cali
Boa Vista
Mitu
Macapa
Quito
Guayaquil
Iquitos
Fonte Boa
Manaus
Santarem
Belem
Sao Luis
Fortaleza
Imperatriz
Teresina
Piura
Natal
Recife
Trujillo
Rio Branco
Porto Velho
Porto Nacional
Lima
Cusco
Aracaju
Salvador
Ica
Arequipa
Trinidad
La Paz
Cochabamba
Santa Cruz
Cuiaba
Brasilia
Goiania
Arica
Sucre
Belo Horizonte
Vitória
Rio de Janeiro
Sao Paulo
Antofagasta
Asuncion
Curitiba
San Miguel de Tucuman
Resistencia
Florianopolis
Porto Alegre
Cordoba
Mendoza
Rosario
Salto
Valparaiso
Buenos Aires
Montevideo
Santiago
Concepcion
Mar del Plata
Bahia Blanca
Valdivia
San Carlos de Bariloche
Comodoro Rivadavia

BBC

British Broadcasting Service

The British Broadcast Service transmits worldwide, 24 hours a day, every day. Next to the Voice of America, it is the most reliable source of news and information on the shorwave bands.

0400 UTC

0400-0430 UTC 17.885 MHz
0400-0600 UTC 7.105 MHz
0400-0630 UTC 3.955 MHz
0400-0730 UTC 7.230 MHz
0430-1730 UTC 21.470 MHz

0500 UTC

0500-0530 UTC 9.600, 17.885 MHz
0500-0545 UTC 6.005 MHz
0500-1800 UTC 6.190 MHz

0600 UTC

0600-0630 UTC 9.600 MHz
0600-0830 UTC 7.150, 17.790 MHz
0600-1030 UTC 17.830 MHz
0600-1400 UTC 17.885 MHz

0700 UTC

0700-0730 UTC 9.600 MHz
0700-1730 UTC 21.660 MHz

0800 UTC

0800-0830 UTC 9.600 MHz
0800-1830 UTC 17.640 MHz

0900 UTC

0900-1100 UTC 15.190 MHz
0900-1500 UTC 6.045 MHz
0900-1730 UTC 17.860 MHz

1100 UTC

1100-1400 UTC 9.515 MHz
1100-1630 UTC 15.205 MHz
1100-1700 UTC 11.750 MHz

1300 UTC

1300-1500 UTC 11.820 MHz
1300-1615 UTC 7.180 MHz
1300-2400 UTC 17.830 MHz

1400 UTC

1400-2130 UTC 17.880 MHz

1500 UTC

1500-1530 UTC 21.490 MHz
1500-1730 UTC 11.775 MHz
1500-2200 UTC 6.195 MHz
1500-2330 UTC 9.410 MHz

1600 UTC

1600-1745 UTC 3.915 MHz
1630-1830 UTC 5.975 MHz
1630-2200 UTC 3.255 MHz

1700 UTC

1700-2030 UTC 7.160 MHz
1700-2200 UTC 6.180 MHz

1800 UTC

1800-2200 UTC 3.955, 7.325, 9.600 MHz

2100 UTC

2100-2400 UTC 9.590 MHz

2200 UTC

2200-0330 UTC 11.750 MHz

2300 UTC

2300-0030 UTC 7.145 MHz
2300-0330 UTC 6.175 MHz

TIME CONVERSION CHART

U.T.C.	PST	PDST MST	MDST CST	CDST EST	EDST
0:00	4 pm	5 pm	6 pm	7 pm	8 pm
1:00	5 pm	6 pm	7 pm	8 pm	9 pm
2:00	6 pm	7 pm	8 pm	9 pm	10 pm
3:00	7 pm	8 pm	9 pm	10 pm	11 pm
4:00	8 pm	9 pm	10 pm	11 pm	Midnight
5:00	9 pm	10 pm	11 pm	Midnight	1 am
6:00	10 pm	11 pm	Midnight	1 am	2 am
7:00	11 pm	Midnight	1 am	2 am	3 am
8:00	Midnight	1 am	2 am	3 am	4 am
9:00	1 am	2 am	3 am	4 am	5 am
10:00	2 am	3 am	4 am	5 am	6 am
11:00	3 am	4 am	5 am	6 am	7 am
12:00	4 am	5 am	6 am	7 am	8 am
13:00	5 am	6 am	7 am	8 am	9 am
14:00	6 am	7 am	8 am	9 am	10 am
15:00	7 am	8 am	9 am	10 am	11 am
16:00	8 am	9 am	10 am	11 am	Noon
17:00	9 am	10 am	11 am	Noon	1 pm
18:00	10 am	11 am	Noon	1 pm	2 pm
19:00	11 am	Noon	1 pm	2 pm	3 pm
20:00	Noon	1 pm	2 pm	3 pm	4 pm
21:00	1 pm	2 pm	3 pm	4 pm	5 pm
22:00	2 pm	3 pm	4 pm	5 pm	6 pm
23:00	3 pm	4 pm	5 pm	6 pm	7 pm

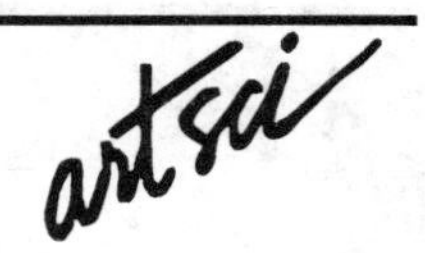

Frequency	Mode	Call Sign	Service / Times	
.1225 MHz	FAX	CFH	WEATHER, HALIFAX, NOVA SCOTIA	
.124	CW	CKN	WEATHER	
.1425	CW	UIK	WEATHER	
.198	AM		BRITISH BROADCASTING CORP. 2300-0445 UTC	
.200			BEACON	Anaheim Lake, BC
.215	CW	CLB	MARKER	
.225	CW	SDL	MARKER	
.231	CW	SOL	MARKER	
.277	CW	ACE	MARKER	
.280	CW	XSD	MARKER	
.283	CW	RSZ	MARKER	
.285	CW	DPG	MARKER	
.292	CW	AOP	MARKER	
.295	CW	BMC	MARKER	
	CW	TOR	MARKER	
.304	CW	"L"	MARKER	
.305	CW	RO	MARKER	
.316	CW	"F"	MARKER	
.326	CW	MCY	MARKER	
.329	CW	TAD	MARKER	
	CW	ROB	MARKER	
.335	CW	STI	MARKER	
.346	CW	XX	MARKER	
.350			BEACON	Raleigh, NC
.353	CW	NO	MARKER	
.354	CW	LWT	MARKER	
.361	CW	BO	MARKER	
.370	CW	SIR	MARKER	
	CW	GYM	MARKER	
.373	CW	ITU	MARKER	
.374	CW	EKG	MARKER	
.375	CW	LV	MARKER	
	CW	VF	MARKER	
.377	CW	HPL	MARKER	
.379	CW	IWW	MARKER	
.382	CW	GC	MARKER	
	CW	BBL	MARKER	
.385	CW	PI	MARKER	
	CW	CRP	MARKER	
.387	CW	WL	MARKER	

Frequency	Mode	Call Sign	Service / Times
.388 MHz	CW	SYF	MARKER
.391	CW	EHY	MARKER
	CW	CMH	MARKER
.394	CW	PNA	MARKER
.396	CW	ZBB	MARKER
.399	CW	LLJ	MARKER
	CW	SB	MARKER
.400	CW	FN	MARKER
.401	CW	SU	MARKER
.402	CW	QQ	MARKER
.403	CW	FN	MARKER
.407	CW	CO	MARKER
.410	CW		INTERNATIONAL MARITIME RADIO DIRECTION FINDING
	CW	DAO	MARKER
.412	CW	RD	MARKER
.413	CW	OER	MARKER
.416	CW	SKX	MARKER
	CW	VOK	WEATHER
	CW	LPO	WEATHER
	CW	L2W	WEATHER
.4165	CW	FJP	WEATHER
.4175	CW	ZLW	WEATHER
.418	CW	ZSC	WEATHER
.420	CW	VFN	WEATHER
	CW	WPP	WEATHER
	CW	UAJ	WEATHER
.4205	CW	VIT	WEATHER
.421	CW	KFS	MARKER
.426	CW	KFS	DE KFS WX SKED 0450/1050/1650/2250 UTC DAILY BT NW QRU ? QSW/AMVERS QSX 500/HF K
.428	CW	WMH	WEATHER
.430	CW	VJZ	WEATHER
	CW	VFF	WEATHER
	CW	VIM	WEATHER
	CW	P2R	WEATHER
.432	CW	NMG	U.S. COAST GUARD
.434	CW	WLO	DE WLO 2 TFC LIST NOW ANS MF/HF K
	CW	VII	WEATHER
.435	CW	OST	MARKER
	CW	VIB	WEATHER

Frequency	Mode	Call Sign	Service / Times
.436 MHz	CW	KFS	DE KFS/A OR KFS/B NW QSX 500/HF QRU ?
.438	CW	CFH	WEATHER
.440	CW	NMA	U.S. COAST GUARD
	CW	NMO	U.S. COAST GUARD
	CW	YIR	WEATHER
	CW	VIS	WEATHER
.444	CW	FUE	MARKER
	CW	SUH	WEATHER
.445	CW	VID	WEATHER
.446	CW	VCF	WEATHER
.448	CW	NMN	U.S. COAST GUARD
	CW	NMC	U.S. COAST GUARD
.450	CW	NOJ	U.S. COAST GUARD
	CW	UBE	WEATHER
.464	CW	2JA	WEATHER
	CW	VCO	WEATHER
.466	CW	NRV	U.S. COAST GUARD
.470	CW	NOJ	U.S. COAST GUARD
.472	CW	NMF	U.S. COAST GUARD
	CW	NMC	WEATHER
	CW	VIA	WEATHER
.474	CW	VCC	WEATHER
.476	CW	9YL	WEATHER
.478	CW	VON	WEATHER
	CW	WNU	WEATHER
	CW	VAE	WEATHER
.482	CW	WSC	U.S. COAST GUARD
.484	CW	KLC	WEATHER
	CW	P2M	WEATHER
	CW	VCS	WEATHER
.4845	CW	8PO	WEATHER
.4875	CW	ZLD	WEATHER
.489	CW	UKB	MARKER
	CW	VAU	WEATHER
	CW	VCM	WEATHER
.500	CW		CALLING AND DISTRESS
.515			BEACON Columbus, OH
	CW	ZLB	WEATHER
.518			NAVTEX
.521	CW	3DP	WEATHER
.522	CW		HOMING, AIR TO AIR, AIR-SURFACE

Frequency	Mode	Call Sign	Service / Times
.5225 MHz	CW	XSG	WEATHER
	CW	ZBP	WEATHER
.530	AM		TRAVELERS INFORMATION SERVICE ACROSS U.S.
.532	CW		HOMING, AIR TO AIR, AIR-SURFACE
.550	AM		FALKLAND ISLAND GOV'T, 0000-2359 UTC
.560	AM		VOICE OF GUYANA (VOG), BROADCASTS FROM 0430 AM, (Guyana Time) to Midnight, except Fridays when sign off is at 0100 AM (Saturday)
.619	CW	ZSV	WEATHER
.621	AM		VOICE OF AMERICA, (English to Africa service) 0300-0430, 1700-2100 UTC (Portuguese to Africa) 1730-1800 UTC
.639	AM		BRITISH BROADCASTING CORP. 0300-0330, 0600-1245 UTC
.648	AM		BRITISH BROADCASTING COPP. 2000-0330, 0500-0530, 0600-0630, 2200-2330 UTC
.675	AM		BRITISH BROADCASTING CORP. 0000-2400 UTC
.690	AM		UNITED NATIONS RADIO, (Spanish service), 1930-2000 UTC
.702	AM		BRITISH BROADCASTING CORP., 2100-0330 UTC
	AM		ISLAMIC REPUBLIC OF IRAN BROADCASTING, (English Service, West Asia), 1130-1230, 1400-1500 UTC
.720	AM		RADIO FREE EUROPE (Polish), 0400-0600, 1800-2200 UTC
	AM		BRITISH BROADCASTING CORP. 0300-0330, 0600-0630 UTC
.738	AM		RADIO FREE EUROPE (Polish), 0400-0600, 1500-2200 UTC
	AM		KOL ISRAEL RUSSIAN SECTION, 0500-0540 UTC

Frequency	Mode	Call Sign	Service / Times
.792 MHz	AM		VOICE OF AMERICA, (English to Middle East/Europe service), 0000-0330, 0400-0430, 0500-0700, 1700-1800, 2230-2400 UTC (Serbo-Croatian to Europe), 0445-0500, 2030-2130 UTC (Romanian to Europe), 0430-0445, 1800-2000 UTC (Ukrainian to the USSR), 0300-0400 UTC
.810	AM		RADIO BERLIN, GERMANY, 1900-2220 UTC
.819	AM		RADIO FREE EUROPE, (Polish), 0400-0600, 1500-2200 UTC
.930	AM		VOICE OF AMERICA, (Service to Caribbean), 0000-0030, 1000-1030, 1100-1200 UTC, (Service to Latin America), 0100-0200 UTC
	AM		BRITISH BROADCASTING CORP., 2000-2400 UTC

TIME CONVERSION CHART

U.T.C.	PST	PDST MST	MDST CST	CDST EST	EDST
0:00	4 pm	5 pm	6 pm	7 pm	8 pm
1:00	5 pm	6 pm	7 pm	8 pm	9 pm
2:00	6 pm	7 pm	8 pm	9 pm	10 pm
3:00	7 pm	8 pm	9 pm	10 pm	11 pm
4:00	8 pm	9 pm	10 pm	11 pm	Midnight
5:00	9 pm	10 pm	11 pm	Midnight	1 am
6:00	10 pm	11 pm	Midnight	1 am	2 am
7:00	11 pm	Midnight	1 am	2 am	3 am
8:00	Midnight	1 am	2 am	3 am	4 am
9:00	1 am	2 am	3 am	4 am	5 am
10:00	2 am	3 am	4 am	5 am	6 am
11:00	3 am	4 am	5 am	6 am	7 am
12:00	4 am	5 am	6 am	7 am	8 am
13:00	5 am	6 am	7 am	8 am	9 am
14:00	6 am	7 am	8 am	9 am	10 am
15:00	7 am	8 am	9 am	10 am	11 am
16:00	8 am	9 am	10 am	11 am	Noon
17:00	9 am	10 am	11 am	Noon	1 pm
18:00	10 am	11 am	Noon	1 pm	2 pm
19:00	11 am	Noon	1 pm	2 pm	3 pm
20:00	Noon	1 pm	2 pm	3 pm	4 pm
21:00	1 pm	2 pm	3 pm	4 pm	5 pm
22:00	2 pm	3 pm	4 pm	5 pm	6 pm
23:00	3 pm	4 pm	5 pm	6 pm	7 pm

Frequency	Mode	Call Sign	Service / Times
1.010 MHz	AM		RADIO "THE VOICE OF VIETNAM", (English Service), 1100-1130
1.080	AM		RADIO FREE EUROPE, (Polish), 0400-0600, 1500-2200 UTC
1.143	AM		VOICE OF AMERICA, (English to Pacific service), 1200-1230, 1400-1500 UTC Vietnamese to Asia) 1230-1330 UTC, (Mandarion to East Asia) 1100-1200, 1330-1400, 1600-1700, 2000-2100 UTC
1.170	AM		RADIO SOUTH KOREA, (KBS), (English Service), 1930-2000 UTC
1.180	AM		VOICE OF AMERICA, (Spanish to Cuba), 0600-0600 UTC
1.188	AM		RADIO DUBLIN LIMITED, English Broadcasts 24 hours daily.
1.197	AM		VOICE OF AMERICA, (English to Middle East/Europe service), 0800-1000, 1700-1730 UTC (English to VOA Europe), 0630-1700 UTC (Sloven to Europe), 0430-0500 UTC (Russian to USSR), 0300-0400 UTC (Polish to Europe), 0530-0630, 2100-2400 UTC (Hungarian to Europe), 1800-2000 UTC
	AM		BRITISH BROADCASTING CORP., 0300-0800, 0900-2200 UTC
1.206	AM		RADIO FREE EUROPE, (Polish), 0400-0600, 1500-2200 UTC
1.224	AM		ISLAMIC REPUBLIC OF IRAN BROADCASTING, (English Service, Middle East), 1130-1230 UTC, (Simulcasts on 702 ,7215 , 9695 UTC)

Frequency	Mode	Call Sign	Service / Times
1.240 MHz	AM		RADIO "THE VOICE OF VIETNAM", (English Service), 1000-1030, 1230-1300, 1330-1400, 2330-2400 UTC, (Spanish Service), 1100-1130 UTC
1.260	AM		VOICE OF AMERICA, (English to Middle East/Europe service) 0100-0330, 1330-1400, 1400-1600, 2100-2200, 2230-2400 UTC (English to VOA Europe), 0300-0330, 0800-1000 UTC (Ukrainian to the USSR), 2000-2100 UTC (Pashto to Middle East), 0000-0030 UTC
1.287	AM		RADIO FREE EUROPE, (Czechoslovakian), 0300-0600, 1000-2200
1.296	AM		BRITISH BROADCASTING CORP., 0300-0330, 0430-0500, 0600-0630, 2200-2330 UTC
1.305	AM		RADIO FREE EUROPE (Polish), 0400-0600, 1500-2200 UTC
1.314	AM		RADIO NORWAY INTERNATIONAL, 1700 UTC
1.323	AM		BRITISH BROADCASTING CORP., 0000-2400 UTC
1.350	AM		VOICE OF AMERICA, (Service to Middle East), 0000-0500, 1330-2200, 2230-2300 UTC
1.413	AM		BRITISH BROADCASTING CORP., 0200-0230, 1300-1400, 1730-1830
1.503	AM		POLISH RADIO WARSAW, (English Service), 1300-1355, 1400-1425, 1500-1525, 1600-1625, 1730-1755, 1900-1925, (Spanish Service), 1430-1455, 2130-2155 UTC
1.512	AM		BRUSSELS CALLING, (Daily), 0730-0030 UTC

Frequency	Mode	Call Sign	Service / Times
1.530 MHz	AM		VOICE OF AMERICA, (English to American Republics service), 0030-0100 UTC. (Spanish to American Republics), 0100-0400, 1200-1430 UTC
1.575	AM		VOICE OF AMERICA, (English to Pacific service) 2230-2400, 0030-0100 UTC. (Lao to East Asia), 1100-1130 UTC (English to VOA Europe), 1530-1600 UTC (Khmer to East Asia), 1400-1500, 2200-2330 UTC
	AM		ARMED FORCES RADIO, JAPAN, U.S. AIR FORCE, 0005-2205 UTC
1.580	AM		VOICE OF AMERICA, (English to Caribbean), 0000-0200, 1000-1200 UTC (English to American Republics service), 0030-0200 UTC (Spanish to American Republics), 0100-0400, 1200-1430 UTC
1.610			TRAVELERS INFORMATION SERVICE ACROSS U.S.
1.647	TONE		LOCATION BEACON
1.716	TONE		LOCATION BEACON
1.720	TONE		LOCATION BEACON
1.727	TONE		LOCATION BEACON
1.731	TONE		LOCATION BEACON
1.762	TONE		LOCATION BEACON
1.800 MHz	**SSB/CW**	**START**	**OF AMATEUR RADIO 160 METER BAND (Ends 20000)**
1.818	CW	W1AW	CW BULLETINS
1.890	LSB	W1AW	ARRL VOICE BULLETINS

Frequency	Mode	Call Sign	Service / Times
2.020 MHz	SSB	VJC	ROYAL FLYING DOCTOR, AUSTRALIA, 0700-1900 UTC
2.065	USB		SHIP TO SHIP, SHIP TO SHORE COMMUNICATIONS.
2.079	USB		SHIP TO SHIP, SHIP TO SHORE COMMUNICATIONS.
2.096	USB		SHIP TO SHIP, SHIP TO SHORE COMMUNICATIONS.
2.130	RTTY/USB		NAVY U.S. COASTAL
2.150	RTTY/USB		NAVY HARBOR CONTROL (US)
2.182	VOICE		INTERNATIONAL DISTRESS AND CALLING
2.260	SSV	VJN	ROYAL FLYING DOCTOR, AUSTRALIA 0800-1700 UTC
2.261	USB		U.S. COAST GUARD AIRCRAFT
2.280	SSB	VKJ	ROYAL FLYING DOCTOR, AUSTRALIA 0700-1700 UTC
2.326	USB		U.S. ARMY MILITARY
2.348	USB		U.S. ARMY MILITARY
2.360	SSV	VKY	ROYAL FLYING DOCTOR, AUSTRALIA 0800-2400 UTC
2.368	RTTY/USB		NAVY HARBOR CONTROL (US)
2.3725	RTTY		CIVIL AIR PATROL
2.374	VOICE		CIVIL AIR PATROL
2.3755	SSB		CIVIL AIR PATROL
2.376	USB/ARQ		COMSAC ALASKA, U.S. COAST GUARD
2.430	USB		INTERNATIONAL SSB RADIO TELEPHONY, (Receive 02572)
2.434	RTTY/USB		NAVY HARBOR CONTROL (US)
2.500	VOICE	WWV	**INTERNATIONAL STANDARDS TIME FREQUENCY**
2.550	RTTY/USB		NAVY HARBOR CONTROL (US)
2.593	ARQ		INTERPOL FREQUENCIES
2.622	USB		NASA MISSION FREQ
2.630	RTTY/USB		NAVY EMERGENCY NET
2.638	USB		U.S. COAST GUARD
2.656	SSB	VJO	ROYAL FLYING DOCTOR, AUSTRALIA 0700-1700 UTC
2.675	USB		U.S. COAST GUARD CUTTERS
2.690	USB		U.S. COAST GUARD ALASKAN WATERS
2.694	USB		U.S. COAST GUARD ATLANTIC WATERS
2.716	USB		NASA MISSION FREQ
	USB		U.S. COAST GUARD CUTTERS

Frequency	Mode	Call Sign	Service / Times
2.732 MHz	RTTY/USB		NAVY SUBS
2.745	FAX		NAVY FAX
2.754	CW		VVV VVV VVV DE CKN CKN CKN CJ3E CJ3E
2.760	SSB	VIO	ROYAL FLYING DOCTOR, AUSTRALIA 0700-1700 UTC
2.764	USB		NASA MISSION FREQ
2.792	SSB	VJB	ROYAL FLYING DOCTOR, AUSTRALIA 0700-1700 UTC
2.8085	USB		U.S. COAST GUARD LAW ENFORCEMENT
2.81385	FAX	GYA	NORTHWOOD, UNITED KINGDOM, 1630-0730 UTC
2.836	RTTY/USB		HARBOR CONTROL
2.840	ARQ		INTERPOL FREQUENCIES
2.8645	USB		AERONAUTICAL WEATHER REPORTING

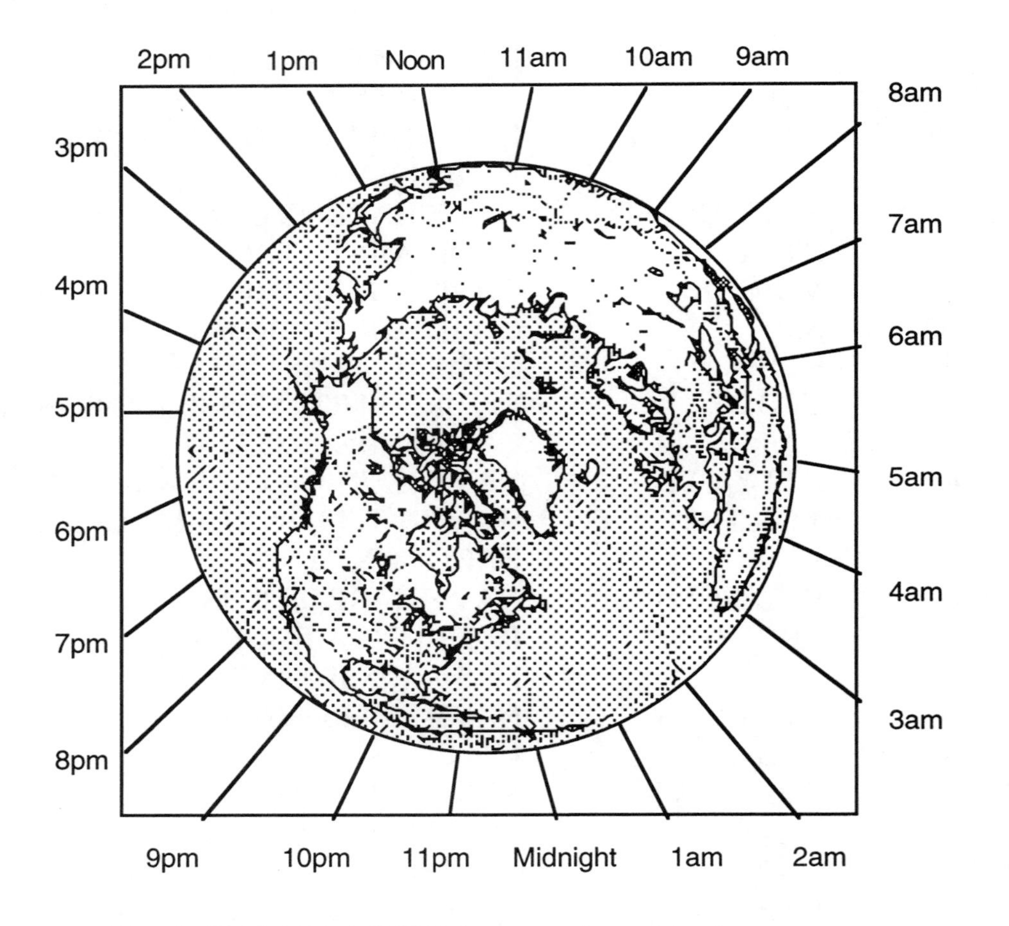

Each "HOUR LINE" is equal to 15 degrees

Frequency	Mode	Call Sign	Service / Times
3.0235MHz	VOICE/CW		INTERNATIONAL HF SEARCH AND RESCUE FREQUENCY, (worldwide search and rescue)
	USB		SHIP TO SHIP, SHIP TO SHORE COMMUNICATIONS.
3.024	RTTY/USB		SEARCH/RESCUE
3.045	VOICE		SPY STATION, SPANISH FEMALE VOICE, SENDS 5-DIGIT TRAFFIC, 0700 UTC
3.050	USB		NAVY AIR/GROUND
3.053	USB		NAVY TACTICAL
3.067	USB		SKYKINGS BROADCASTS
3.095	USB		NAVY TACTICAL
3.109	USB		NAVY AIR/GROUND
3.113	USB		SKYKING BROADCASTS
	USB		LOOKING GLASS
3.123	USB		U.S. COAST GUARD AIRCRAFT
3.130	USB		COMSTA PACIFIC
3.160	USB	MKG	R.A.F., LONDON
3.180	CW		X HF BEACON LOCATED IN PRAGUE
3.187	USB		NASA MISSION FREQ
3.220	AM	HCJB	THE VOICE OF THE ANDES (Spanish Service) 0200-0500
3.228	CW		CKN NAWS DE CKN II ZKR F1 2386 4170 6251 8324 12386 16568 22191 KHZ AR DE CKN
3.235	FAX	VFF	FROBISHER, CANADA
3.240	AM		RADIO SWITZERLAND, 0300-0355 UTC
3.242	FAX		NIK BOSTON, MASS.
3.253	USB		U.S. COAST GUARD 1ST DISTRICT
	FAX	VFF	WEATHER, FRIBISHER, CANADA
3.255	AM		BRITISH BROADCASTING CORP., 0300-0530, 1630-2200 UTC
3.261	USB		NORAD
3.265	USB		NAVY TACTICAL
3.268	CW	NPG	U.S. COAST GUARD
3.2685	USB		U.S. ARMY MILITARY
3.275	USB/RTTY		U.S. ARMY MARS
3.2865	CW	CKN	NAWS DE CKN II ZKR F1 2386 4152.6 6238.6 8333.5 12478.5 16642 22178 KHZ AR
3.2895MHz	FAX	GFA	BRACKNELL, UNITED KINGDOM,

Frequency	Mode	Call Sign	Service / Times
			0000-2400 UTC
3.290	RTTY	P	HF BEACON LOCATED IN KALININGRAD, (Will send traffic on RTTY at 75 baudot)
3.292	USB		SKYKING BROADCASTS
3.295	USB		SKYKING BROADCASTS
	USB		LOOKING GLASS
3.310	ARQ	KRH50	U.S. DEPT. OF STATE, AZORES
3.330	VOICE	CHU	OTTAWA CANADA, STANDARD TIME/FREQUENCY
3.3495	CW		U.S. ARMY MARS, AAA6USA
3.350	CW		U.S. ARMY MARS
3.356	RTTY		EGYPT NEWS AGENCY
3.357	FAX	NAM	NORFOLK, VA, WEATHER
3.365	AM		RADIO GUATEMALA, 0200-0400 UTC
3.369	USB		SKYKING BROADCASTS
3.407	USB		NATIONAL HURRICANE CENTER
3.413	USB		SAN FRANCISO INTERNATIONAL AIR TRAFFIC, (Secondary) (Primary 05547 USB)
3.4365	FAX	GYA	NORTHWOOD, UNITED KINGDOM APRIL 1 THRU SEPTEMBER 29, 1930-0400 UTC SEPTENBER 30 THRU MARCH 31, 1530-0830 UTC

3.500 MHz SSB/CW START OF AMATEUR RADIO 80 METER BAND (Ends 4.000)

Frequency	Mode	Call Sign	Service / Times
3.585	CW	W1AW	CW BULLETINS
3.593	LSB		INTERNATIONAL PACKET NET
	ARQ		INTERPOL FREQUENCIES
3.625	RTTY		ARRL RTTY BULLETINS
3.705	ARQ		INTERPOL FREQUENCIES
3.714	ARQ		INTERPOL FREQUENCIES
3.717	ARQ		INTERPOL FRENQUENCIES, ZURICH
3.762	RTTY		JORDAN NEWS AGENCY
3.801	ARQ		INTERPOL FREQUENCIES
3.840	LSB		AMATEUR AMSAT NET
3.845	LSB		SLOW SCAN TELEVISION
3.858	FAX	DDH	WEATHER, QUICKBORN, GERMANY
3.915 MHz	AM		BRITISH BROADCASTING CORP.,

Frequency	Mode	Call Sign	Service / Times
			1600-1745 UTC
3.927	VOICE		SPY STATION, SPANISH FEMALE VOICE, SENDS 5-DIGIT TRAFFIC, 0200, 0300, 0400, 0500 UTC
	CW		SPY STATION, MORSE CODE, SENDS 5-DIGIT TRAFFIC, 0130, 0135, 0300, 0330, 0400 UTC
3.955	AM		BRITISH BROADCASTING CORP., 0400-0630, 1800-2200 UTC
3.980	AM		VOICE OF AMERICA, (English to Middle East/Europe service), 0630-0700, 1700-1730 UTC (Sloven to Europe), 0430-0500 UTC (Polish to Europe), 0530-0630, 2100-2200, 2200-2400 UTC
3.985	AM		RADIO FREE EUROPE, (Latvian), 0200-0300 UTC
	AM		SWISS RADIO INTERNATIONAL, 0730-0800 UTC
3.990	LS	W1AW	ARRL VOICE BULLETIN
3.995	AM		RADIO BERLIN, GERMANY, 0600 UTC

DATE	FREQ.	MODE	TIME	STATION	SIGNAL	COMMENTS	SWL SENT	SWL REC'D

Frequency	Mode	Call Sign	Service / Times
4.000 MHz	USB		MILITARY MARS
4.0015	CW		NAVY MARS
4.0035	USB		AIR FORCE MARS
4.005	USB	NAM	NAVAN COMM AREA MASTER STATION, NORFOLK, VA
4.007	USB		NAVY MARS
4.008	USB		NAV-8 NAVY MARS, HONOLULU, HI
4.010	RTY	NAV	NAVCOMMSTA, SAN JUAN, PR
	SSB	VJI	ROYAL FLYING DOCTOR, AUSTRALIA, 0800-1700 UTC
4.011	USB		NAVY MARS
4.014	USB		NAVY AIR/GROUND
	FAX	ZRO	WEATHER, PRETORIA, REPUBLIC OF SOUTH AFRICA
4.015	RTTY	NMH	USCG, ALEXANDRIA, VA
4.0185	USB	WAR	HF/MARS RADIO STATION, FT DETRICK, MD
4.020	USB		ARMY MARS
	VOICE		SPY STATION, SPANISH FEMALE VOICE, SENDS 5-DIGIT TRAFFIC, 0400 UTC
	CW		SPY STATION, MORSE CODE, SENDS 5-DIGIT TRAFFIC, 0500 UTC
4.025	USB		ARMY MARS
4.028	VOICE		SPY STATION, SPANISH FEMALE VOICE, SENDS 5-DIGIT TRAFFIC, 0500, 0555, 0600, 0700, 0705, 1000 UTC
	CW		SPY STATION, MORSE CODE, SENDS 5-DIGIT TRAFFIC, 0000, 0300, 0400, 0430, 0900 UTC (WILL REPEAT TRAFFIC SEND ON 5.854)
4.030	USB		AAE HF/MARS RADIO STATION, FT LEWIS, WA
	SSB	VKL	ROYAL FLYING DOCTOR, AUSTRALIA, 0700-1700 UTC
4.035	USB		ARMY MARS
4.0375	CW		CUBA MILITARY
	FAX	SMA	WEATHER, SWEDEN
4.038	LSB		ARMY MARS
4.040	RTTY	NPG	USN COMMSTA, SAN FRANCISCO, CA
4.041	USB		NAVY MARS
	USB		ARMY MARS

4

Frequency	Mode	Call Sign	Service / Times
4.044 MHz	CW		SPY STATION, MORSE CODE, SENDS 5-DIGIT TRAFFIC, 0330 UTC
4.045	USB		NORAD
	SSB	VJJ	ROYAL FLYING DOCTOR, AUSTRALIA, 0800-1700 UTC
4.053	FAX	AOK	WEATHER, ROTA, SPAIN
4.055	SSB	VJC	ROYAL FLYING DOCTOR, AUSTRALIA, 0700-1900 UTC
4.0565	RTY	TJK	RY'S 50 BAUD, BAUDOT
4.063	USB		NAVY MARS
4.077	USB		INTERNATIONAL SSB RADIO TELEPHONY, (Receive on 4369)
4.104	USB		INTERNATIONAL SSB RADIO TELEPHONY, (Receive on 4396)
4.119	USB		INTERNATIONAL SSB RADIO TELEPHONY, (Receive on 4411)
4.125	USB		MARINE SHIP TO SHIP
	USB/CW		CALLING AND DISTRESS, (Secondary)
4.1435	USB		NAVY FLEET ACTIVE
4.1436	USB		SHIP TO SHIP, SHIP TO SHORE COMMUNICATIONS.
4.1492	USB		MARINE SHIP TO SHIP
4.152	USB		NAVY TACTICAL
4.125	USB		MARINE WEATHER FORECAST
4.177	USB		JAPANESE MARITIME AGENCY
4.1815	USB		JAPANESE MARITIME AGENCY
4.200	CW	BKXC	XFV DE BKXC, (Possible chinese military cw net) 0520 UTC
4.2015	ARQ/CW		COMSAC ALASKA, U.S. COAST GUARD
4.2115	CW	KFS	DE KFS KFS KFS SITOR SELCAL 11.094 QSX 4 6 8 12 AND 16 MHZ
4.2117	ARQ	GKE2	
4.2125	ARQ	WLO	(Receive on 041745)
4.213	ARQ	WLO	(Receive on 04175)
4.214	ARQ	ZSC	DE ZSC SITOR SVC K
4.2145	ARQ	KLB	
4.215	ARQ	WLO	(Receive on 04177)
4.2164	ARQ	HEC	
	ARQ	KPH	
4.2165	ARQ	WCC	
4.217	ARQ	WLO	(Receive on 041795)

Frequency	Mode	Call Sign	Service / Times
4.2185MHz	ARQ	WLO	
4.223	FAX	NMF	COMMSTA, BOSTON, MA
	FAX	KWX	WEATHER, LEWES, DELWARE
4.2267	CW	XFM	CQ CQ DE XFM XFM QRU 4/8/12
4.2305	CW	VIB	WEATHER/AMVER
4.2325	CW	FUF	VVV DE FUF
4.2355	CW	VAI	CQ CQ CQ DE VAI VAI VAI QSX 4 /6/8/12/16 MHZ CH7/CH8 CH3/CH4 22 MHZ ON REQUEST OBS/AMVER/QRJ ? VAI SITOR SELCALL 1 581 QRU ? K
4.2466	CW	ZRH	DE ZRH QSX 4 6 8 12 XX K
4.2465	CW	KPH	CQ CQ DE KPH KPH KPH QSX 4 6 8 12 16 22 MHZ K
4.247	CW	VJZ	WEATHER/AMVER
4.24785	FAX	GYA	NORTHWOOD, UNITED KINGDOM, 0000-2400 UTC
	CW	P2R	WEATHER/AMVER
4.2486	CW	ZRQ	VVV VVV VVV ZRQ 2/4
4.2507	CW	XFL	CQ CQ CQ DE XFL XFL XFL
	CW	PPJ	WEATHER/AMVER
4.2523	CW	GKC	DE GKC
4.2554	CW	CFH	VVV VVV VVV DE CFH CFH CFH C13L C13L C13L
	CW	VIT	WEATHER/AMVER
4.256	CW	KLC	WEATHER/AMVER, (Galveston, Texas)
4.2575	CW	WLO	DE WLO 2 OBS ? AMVERS ? QSX 4 6 8 12 16 22 25R071 MHZ NW ANS C9 K CQ CQ CQ DE WLO WLO QSW 434 4257R5 446R5 8445R5 8473R5 8758R0 12660R0 12774R5 13024R0 16968R5 17172R4 AND 2676R5 KHZ TFC LIST AND WX QSW 4343 8514 12776R5 17022R5 AND 22487 KHZ EVERY HOUR ON THE HOUR..WARC BT 87 FREQUENCY CHANGE EFFECTIVE 0001 UTC JULY 1 1991 CWS INFO 0200 0700 000 GMT FOLLOWING TFC LIST FEC INFO 0225 0735 2035 GMT FOLLOWING QTC LIST 4343.0 6416.0 8 8514.0 12786.5 17022.5 22487.0 KHZ CQ CQ CQ DE WLO WLO

Frequency	Mode	Call Sign	Service / Times
4.265 MHz	CW	PPL	WEATHER/AMVER
4.2675	CW	CKN	VVV VVV VVV DE CKN CKN CKN C13E C13E C13E
4.270	USB	PCD	ISRAELI MOSSAD INTELLIGENCE
4.271	FAX	CFH	HALIFAX, NOVA SCOTIA
	CW	UBE	WEATHER/AMVER
4.2725	CW	VIA	WEATHER/AMVER
	CW	VID	WEATHER/AMVER
4.2734	CW	KFH	CQ DE KFH KFH KFH/B QSX 8 12 16 22 MHZ K
4.275	CW	HPP	VVV VVV CQ CQ CQ DE HPP HPP HPP QTC ? AMVER ? OBS ? QSW RTG 500 KHZ / 4.275 8.589 / 12.699 / 17.232.4 MHZ CH 5/6/8 AND RTTY 12.583.0 / 12.480.5 16.822.5 16.699.5 MHZ
4.2855	CW	VCS	VVV VVV VVV CQ DE VCS VCS VCS QSX 4 6 AND 8 MHZ CHNL 3/4/7/8
4.289	CW	PWZ	WEATHER/AMVER
4.291	CW	ZSC	WEATHER/AMVER
4.2945	CW	WNU31	CQ CQ CQ DE WNU31 WNU31 WNU31 QSX 4 6 8 12 16 MHZ OBS ?
4.296	FAX	NOJ	KODIAK, AK, WEATHER
4.298	CW	PPO	WEATHER/AMVER
4.304	CW	L2X	WEATHER/AMVER
4.3065	CW	LPO	WEATHER/AMVER
4.3095	CW	WNU31	CQ CQ CQ DE WNU41 WNU41 WNU41 QSX 4 6 8 12 16 MHZ OBS ?
4.310	CW	NPM	NAVCOMMSTA, PEARL HARBOR, HI
4.319	CW	NMG	USCG, NEW ORLEANS, LA
4.3205	CW	PPO	CQ CQ CQ DE PPO PPO PPO AS
4.3289	CW	FFL	CQ CQ DE FFL2/FFL3/FFL4 PSE QSX 6270/6278.5 AND 8360/ 8469 KHZ ARK
4.334	CW	PJC	CQ CQ DE PJC PJC TFC LIST COMING ON 4334 AND 8694 KHZ AS
4.340	CW		SUBMARINE DISTRESS CALLING FREQUENCY, NATO
4.343	FEC	WLO	WEATHER REPORTS
	CW	WLO	DE WLO
4.344	FAX		NATIONAL WEATHER SERVICE
4.345	FAX	RCH	RUSSIAN WEATHER
4.346	FAX	NMC	COMMSTA, SAN FRANCISCO, CA

Frequency	Mode	Call Sign	Service / Times
4.3493MHz	CW	KLB	CQ CQ CQ DE KLB KLB QSX 4 6 8 12 16 AND 22 OBS ? SITOR SELCAL 1113 AR K
4.350	ARQ	KFS	
	SSB	VJD	ROYAL FLYING DOCTOR, AUSTRALIA, 0700-1700 UTC
4.3505	ARQ	GKE2	
	ARQ	WNU	
4.351	ARQ	NMN	COMMSTA, PORTSMOUTH, VA
4.3515	ARQ	NMO	
4.353	ARQ	JFA	
4.3535	ARQ	HPP	
	ARQ	ZSC	
4.354	ARQ	VIP37	
4.355	ARQ	NRV	COMMSTA, GUAM
4.3555	ARQ	HPP	
4.356	ARQ	WCC	
4.3565	ARQ	KPH	
	ARQ	WLO	
4.3605	USB	NMN	COMMSTA, NORFOLK, VA
4.369	USB	WLO	WEATHER
4.373	USB		SKYKING BROADCASTS
4.377	USB		U.S. NAVY FLEET TACTICAL
4.3835	LSB		ALASKA STATE EMERGENCY COMMUNICATIONS NET
4.395	VOICE		SPY STATION, GERMAN FEMALE VOICE, SENDS 5-DIGIT TRAFFIC, 2100 UTC
4.4015	CW	KMS	CQ CQ CQ DE KMS KMS KMS
4.402	USB		U.S. NAVY FLEET
4.416 MHz	USB		PACIFIC FLEET
4.4194	USB		SHIP TO SHIP, SHIP TO SHORE COMMUNICATIONS.
4.421	USB	WOO	AT&T SERVICES
4.426	USB		PACIFIC WEATHER REPORTING, NATIONAL WEATHER SERVICE
4.444	ARQ		INTERPOL FREQUENCIES
4.445	CW	NPO	U.S. COAST GUARD
4.4455	USB		CUBA MILITARY
4.447	CW		HF BEACON LOCATED IN MURMANSK, USSR
4.449	USB		USAF TACTICAL AIR COMMAND
4.4645	LSB		CIVIL AIR PATROL
	RTY	TJK	RY'S, 50 BAUD, BAUDOT

Frequency	Mode	Call Sign	Service / Times
4.4647MHz	ARQ	WLO	AMTOR 100 BAUD (NOAA) NON-HURRICANE SEASON 4343.0, 6416.0 8514.0, 12886.5, 1 7021.6, 22487.0, HURRICANE SEASON 4462.5, 6344.0, 8534.0, 12992.0, 16997.6, 22569.0 TRAFFIC LISTS.. FEC MODE HH PLUS 35 WEATHER NOAA.. AFTER TRAFFIC LIST NORTH ATL SEAS 0035 0635 1235 1835 UTC NORTH PAC SEAS 0135 0735 1335 1935 UTC W CENTRAL N ATL O335 0935 1530 UTC
4.466	SSB		CIVIL AIR PATROL
4.492	USB		SKYKING BROADCASTS
	ARQ		INTERPOL FREQUENCIES
4.495	USB		SKYKING BROADCASTS
	USB		LOOKING GLASS
4.5045	USB		CIVIL AIR PATROL
4.5075	USB		CIVIL AIR PATROL
4.526	FAX	SUU	WEATHER, CAIRO, EGYPT
4.5285	RTTY	RUM	RUSSIAN WEATHER
4.555	AM		VOICE OF AMERICA
4.560	USB	YHF	ISRAELI MOSSAD INTELLIGENCE
4.565	AM		VOICE OF AMERICA
4.575	USB		NAVY MARS
4.5835	USB		CIVIL AIR PATROL
4.585	USB		CIVIL AIR PATROL
4.5865	SSB		CIVIL AIR PATROL
4.589	ARQ	KRH50	U.S. DEPT. OF STATE
4.5935	USB		NAVY MARS
4.610	FAX	GFA	BRTACKNELL, UNITED KINGDOM, 1800-2400 UTC
4.623	CW	NGR	U.S. COAST GUARD
4.626	AQR	KRH50	U.S. DEPT. OF STATE
4.627	USB		CIVIL AIR PATROL
4.630	USB		CIVIL AIR PATROL
4.632	ARQ		INTERPOL FREQUENCIES
4.6325	LSB		100 BAUD ARQ, INTERPOL, ZURICH

Frequency	Mode	Call Sign	Service / Times
4.635 MHz	SSB	VJC	ROYAL FLYING DOCTOR, AUSTRALIA, 0700-1900 UTC
4.640	VOICE		SPY STATION, ENGLISH FEMALE VOICE, SENDS 3-2 DIGIT TRAFFIC, 0000 UTC
4.700	USB		US MILITARY NET W/VOICE CHECK IN
4.704	RTTY/USB		U.S. NAVY FLEET TACTICAL
	FAX	AOK	ROTA, SPAIN, WEATHER
4.7105	RTTY/USB		AIR/GROUND TACTICAL
4.725	USB		SKYKING BROADCASTS PRIMARY AIR/GROUND/REFUEL
	USB		LOOKING GLASS, SIERRA 390
4.721	USB		USAF GLOBAL CONTROL AND COMMAND
4.742	USB		USAF TACTICAL AIR COMMAND
4.746	USB		USAF GLOBAL CONTROL AND COMMAND
4.756	ARQ		INTERPOL FREQUENCIES
4.763	USB	NKW	USN, DIEGO GARCIA
4.7775	FAX	IMB	WEATHER, ROME, ITALY
4.782	FAX	GFE	BRACKNELL, UNITED KINGDOM
4.790	AM		RADIO PERU, 0200-0300 UTC
4.801	ARQ		INTERPOL FREQUENCIES
4.820	AM		RADIORADIO LA VOZ EVANGELICA, HONDURAS, 0300-0900 UTC
4.835	AM		RADIO GUATEMALA, 0200-0400 UTC
4.837	ARQ		INTERPOL FREQUENCIES
4.8375	ARQ		WORLD WIDE INTERPOL
4.8391	CW	NAM	NORFOLK, VA
4.853	FAX	NPM	PEARL HARBOR, HI, WEATHER
4.855	CW	NPM	PEARL HARBOR, HI
	ARQ		INTERPOL FREQUENCIES
4.860	SSB	VZK	ROYAL FLYING DOCTOR, AUSTRALIA, 0800-1600 UTC
4.865	USB		OVERSEAS VOICE TRAFFIC
4.880	ARQ		DEPARTMENT OF STATE IN WASHINGTON, D.C.
	SSB	VJN	ROYAL FLYING DOCTOR, AUSTRALIA, 0800-1700 UTC
	USB	ULX	ISRAELI MOSSAD INTELLIGENCE
4.890	CW	DLYD	RTI DE DLYD QRU ? QSA ? NIL K (0450 UTC)
4.895	USB		U.S. ARMY MILITARY
4.896	USB		SKYKING BROADCASTS

Frequency	Mode	Call Sign	Service / Times
4.900 MHz	USB		NASA MISSION FREQ
4.926	SSB	VZK	ROYAL FLYING DOCTOR, AUSTRALIA, 0800-1700 UTC
4.940	SSB	VIO	ROYAL FLYING DOCTOR, AUSTRALIA, 0800-1700 UTC
4.957	CW	KKN39	QRA QRA QRA DE KKN39 KKN39 KKN39 QSX 4/13/17 K QRA QRA QRA DE KKN39 KKN39 KKN39 QSX 4/13/17/25 K (Will transmit call once every minute)
4.970	AM		RADIO RUMBOS, VENEZUELA, 0300-0600 UTC
4.902	USB		NASA MISSION FREQ
4.980	AM		RADIO VENEZUELA, 0300-0500 UTC
	SSB	VJJ	ROYAL FLYING DOCTOR, AUSTRALIA, 0800-1700 UTC
4.992	USB		NASA MISSION FREQ

TIME CONVERSION CHART

U.T.C.	PST	PDST MST	MDST CST	CDST EST	EDST
0:00	4 pm	5 pm	6 pm	7 pm	8 pm
1:00	5 pm	6 pm	7 pm	8 pm	9 pm
2:00	6 pm	7 pm	8 pm	9 pm	10 pm
3:00	7 pm	8 pm	9 pm	10 pm	11 pm
4:00	8 pm	9 pm	10 pm	11 pm	Midnight
5:00	9 pm	10 pm	11 pm	Midnight	1 am
6:00	10 pm	11 pm	Midnight	1 am	2 am
7:00	11 pm	Midnight	1 am	2 am	3 am
8:00	Midnight	1 am	2 am	3 am	4 am
9:00	1 am	2 am	3 am	4 am	5 am
10:00	2 am	3 am	4 am	5 am	6 am
11:00	3 am	4 am	5 am	6 am	7 am
12:00	4 am	5 am	6 am	7 am	8 am
13:00	5 am	6 am	7 am	8 am	9 am
14:00	6 am	7 am	8 am	9 am	10 am
15:00	7 am	8 am	9 am	10 am	11 am
16:00	8 am	9 am	10 am	11 am	Noon
17:00	9 am	10 am	11 am	Noon	1 pm
18:00	10 am	11 am	Noon	1 pm	2 pm
19:00	11 am	Noon	1 pm	2 pm	3 pm
20:00	Noon	1 pm	2 pm	3 pm	4 pm
21:00	1 pm	2 pm	3 pm	4 pm	5 pm
22:00	2 pm	3 pm	4 pm	5 pm	6 pm
23:00	3 pm	4 pm	5 pm	6 pm	7 pm

(C) N6MQS

Frequency	Mode	Call Sign	Service / Times
5.000 MHz	**VOICE**	**WWV**	**INTERNATIONAL STANDARDS TIME AND FREQUENCY**
	VOICE	JJY	TIMING SIGNALS FROM JAPAN
5.010	SSB	VJO	ROYAL FLYING DOCTOR, AUSTRALIA, 0700-1700 UTC
5.011	USB		U.S. ARMY, MILITARY
5.0125	USB		U.S. ARMY, MILITARY
5.020	USB		SKYKING BROADCASTS
5.025	AM		RADIO CUBA, 0100-0800 UTC
5.026	USB		SKYKING BROADCASTS
5.027	RTTY		KUWAIT NEWS SERVICE
5.040	AM		RADIO ECUADOR, 0000-0500 UTC
5.055	AM		RADIO COSTA RICA, 0000-0600 UTC
5.063	LORAN		PACIFIC LORAN STATION
5.0715	CW	NAM	NORFOLK, VA
5.093	FAX	LZJ	WEATHER, SOFIA, BULGARIA
5.095	AM		RADIO PERU, 0600-1100 UTC
5.0974	CW	CFH	NAWS DE CFH II ZKR F1 3287 4156.6 6234.6 8338.5 12449.5 16602.4 22142 KHZ AR ((Halifax, NS))
5.098	FAX	AXM	CANBERRA, AUSTRALIA, WEATHER, 0000-2400 UTC
5.0985	RTTY	JAE	KYODO PRESS, TOKYO NEWS AGENCY, (JAE58/JAT28)
5.104	ARQ		INTERPOL FREQUENCIES
5.107	RTTY		IRAN NEWS AGENCY
5.110	USB		SKYKING BROADCASTS
	SSB	VJI	ROYAL FLYING DOCTOR, AUSTRALIA, 0800-1700 UTC
5.145	SSB	VJN	ROYAL FLYING DOCTOR, AUSTRALIA, 0800-1700 UTC
5.157	USB	MKG	R.A.F.,LONDON
5.161	RTTY	CLN	HAVANA, CUBA
5.1675	USB		ALASKA EMERGENCY USE ONLY, (Amateur Radio Operations)
5.180	USB		NASA MISSION FREQ
5.185	FAX	LRO	WEATHER, BUENOS AIRES, ARGENTINE

Frequency	Mode	Call Sign	Service / Times
5.190 MHz	USB		NASA MISSION FREQ
5.200	SSB	VJB	ROYAL FLYING DOCTOR, AUSTRALIA, 0700-1600 UTC
5.208	RTTY		INTERPOL LONDON, ENGLAND
	ARQ		INTERPOL FREQUENCIES
5.215	USB		SKYKING BROADCASTS
5.2165	RTTY	FUM	FRENCH NAVY
5.220	AM		RADIO EGYPT
	USB		CUBAN MILITARY
5.227	SSB	VJJ	ROYAL FLYING DOCTOR, AUSTRALIA, 0800-1700 UTC
5.230	SSB	VJT	ROYAL FLYING DOCTOR, AUSTRALIA, 0700-1700 UTC
5.243	USB		SKYKING BROADCASTS
5.260	SSB	VKJ	ROYAL FLYING DOCTOR, AUSTRALIA, 0600-1700 UTC
5.272	USB		U.S. COAST GUARD
5.2815	CW	CAL	CHILE WEATHER, (Will send 5 digit letter traffic)
5.297	USB		USAF NORAD HQ.
5.300	SSB	VZK	ROYAL FLYING DOCTOR, AUSTRALIA, 0800-1700 UTC
5.305	ARQ		INTERPOL FREQUENCIES
5.320	USB		U.S. COAST GUARD
5.328	USB		SKYKING BROADCASTS
5.340	SSB	VZZ	ROYAL FLYING DOCTOR, AUSTRALIA, 0800-1600 UTC
5.350	USB		NASA MISSION FREQ
5.355	SSB	VIH	ROYAL FLYING DOCTOR, AUSTRALIA, 0700-1700 UTC
5.360	SSB	VJO	ROYAL FLYING DOCTOR, AUSTRALIA, 0700-1700 UTC
5.370	SSB	VZZ	ROYAL FLYING DOCTOR, AUSTRALIA, 0800-1600 UTC
5.379	CW	KRH50	QRA QRA QRA DE KRH50 KRH50 KRH50 QSX 5/7/11/13/16/20 ? QSX 5/7/11/13/16/20 K (Will transmit call once every minute)
5.400	USB		U.S. ARMY, MILITARY
5.410	SSB	VJD	ROYAL FLYING DOCTOR, AUSTRALIA, 0700-1700 UTC

Frequency	Mode	Call Sign	Service / Times
5.415 MHz	VOICE		SPY STATION, SPANISH FEMALE VOICE, SENDS 5-DIGIT TRAFFIC, 0500 UTC
5.417	VOICE		SPY STATION, SPANISH FEMALE VOICE, SENDS 5-DIGIT TRAFFIC, 0700-0730 UTC
5.427	ARQ	KRH50	U.S. DEPT OF STATE
5.434	AM		VOICE OF AMERICA
5.438	USB	ART	ISRAELI MOSSAD INTELLIGENCE
5.439	USB	DRE	ISRAELI MOSSAD INTELLIGENCE
5.441	VOICE		SPY STATION, SPANISH FEMALE VOICE, SENDS 5-DIGIT TRAFFIC, 0630 UTC
5.445	SSB	VJI	ROYAL FLYING DOCTOR, AUSTRALIA, 0800-1700 UTC
5.446	USB		USMC TACTICAL
5.500	VOICE		SPY STATION, FRENCH FEMALE VOICE, SENDS 5-DIGIT TRAFFIC, 2100 UTC
5.523	FAX	BAF	BEIJING, PEOPLES REPUBLIC OF CHINA, WEATHER
5.562	USB		NATIONAL HURRICANE CENTER
5.547	USB		SAN FRANCISO INTERNATIONAL AIR TRAFFIC (Primary), (Secondary 03413 USB)
5.571	USB		U.S. COAST GUARD JOINT OPERATIONS
5.598	USB		INTERNATIONAL AIRLINES
5.616	USB		INTERNATIONAL AIRLINES
5.64995	USB		US MILITARY NET
5.680	USB		SEARCH/RESCUE, (worldwide search and rescue)
	VOICE/CW		INTERNATIONAL HF SEARCH AND RESCUE FREQUENCY
	USB		U.S. COAST GUARD
	USB		SHIP TO SHIP, SHIP TO SHORE COMMUNICATIONS.
5.684	USB		SKYKING BROADCASTS
5.692	USB		U.S. COAST GUARD AIRCRAFT
	USB		U.S. AIR FORCE
5.696	USB		U.S. COAST GUARD AIRCRAFT
5.700	USB		SKYKING BROADCASTS
	USB		LOOKING GLASS
5.703	USB		USAF MIDDLE EAST
	USB		US PILOT'S TACTICAL
5.718	RTTY/USB		FLEET TACTICAL

Frequency	Mode	Call Sign	Service / Times
5.725 MHz	VOICE		SPY STATION, ENGLISH FEMALE VOICE, SENDS 5-DIGIT TRAFFIC, 0405 UTC
5.731	SSB	VZK	ROYAL FLYING DOCTOR, AUSTRALIA, 0800-1600 UTC
5.735	SSB	VJC	ROYAL FLYING DOCTOR, AUSTRALIA, 0700-1900 UTC
5.740	SSB	VZK	ROYAL FLYING DOCTOR, AUSTRALIA, 0700-1700 UTC
5.745	AM		VOICE OF AMERICA
5.749	VOICE		SPY STATION, ENGLISH FEMALE VOICE, SENDS 5-DIGIT TRAFFIC, 0530 UTC
5.753	FAX	AXI	DARWIN, AUSTRALIA, WEATHER, 0800-2100 UTC
5.758	VOICE		SPY STATION, SPANISH FEMALE VOICE, SENDS 5-DIGIT TRAFFIC, 0200 UTC
	CW		SPY STATION, MORSE CODE, SENDS 5-DIGIT TRAFFIC, 0300 UTC (WILL REPEAT TRAFFIC AT 0330 UTC ON 5.800)
5.762	VOICE		SPY STATION, SPANISH FEMALE VOICE, SENDS 5-DIGIT TRAFFIC, 0200, 0230, 0600 UTC
	VOICE		SPY STATION, MORSE CODE, SENDS 5-DIGIT TRAFFIC, 0130 UTC
5.768	FAX	JBK3	TOKYO, JAPAN, WEATHER, PRESS
5.7835	FAX	NAX	U.S. NAVY
5.790	USB		MEXICAN MILITARY
5.800	USB		LOOKING GLASS, WHISKEY 101
	USB		U.S. COAST GUARD LAW ENFORCEMENT
	FAX	YZZ	WEATHER, BELGRADE, YUGOSLAVIA
	CW		SPY STATION, MORSE CODE, SENDS 5-DIGIT TRAFFIC, 0300, 0430 UTC (WILL TRANSMIT ON 5.758 AT SAME TIME)
5.805	CW		SPY STATION, MORSE CODE, SENDS 5-DIGIT TRAFFIC, 0530 UTC
5.8055	ARQ		SAUDI ARABIA EMBASSY IN CANADA
5.806	CW	ZKLF	(Sends Meteorological reports at 0500, 1800 UTC, report will be in form of 5 digit number traffic, FAX at 0545 UTC) Also transmits on 16339 at same time) CQ CQ CQ DE ZKLF ZKLF ZKLF CQ CQ CQ DE ZKLF ZKLF ZKLF BT

Frequency	Mode	Call Sign	Service / Times
5.807 MHz	FAX	ZKLF	WELLINGTON, NEW ZEALAND, WEATHER, 0000-2400 UTC
5.810	USB		NASA MISSION FREQ
5.812	VOICE		SPY STATION, SPANISH FEMALE VOICE, SENDS 4-DIGIT TRAFFIC, 0400, 0430 UTC (BROADCAST ON 11.532 AT SAME TIME)
5.820	USB		IRAQ MILITARY ARMY NET
	VOICE		SPY STATION, SPANISH FEMALE VOICE, SENDS 5-DIGIT TRAFFIC, 0630 UTC
5.826	USB		SKYKING BRAODCASTS
	USB		LOOKING GLASS
5.835	CW		SPY STATION, MORSE CODE, SENDS 5-DIGIT TRAFFIC, 0400, 0420, 0430, 0500 UTC
5.845	SSB	VNZ	ROYAL FLYING DOCTOR, AUSTRALIA, 0800-1700 UTC
5.849	RTTY		EGYPT NEWS AGENCY
	CW		A1 DE A2
5.850	VOICE		SPY STATION, SPANISH FEMALE VOICE, SENDS 5-DIGIT TRAFFIC, 0500, 0530, 0600 UTC
	SSB	VJN	ROYAL FLYING DOCTOR AUSTRALIA, 0700-1700 UTC
5.852	USB		SECURE MILITARY VOICE NET
5.854	USB	984	U.S. NAVY
	CW		SPY STATION, MORSE CODE, SENDS 5-DIGIT TRAFFIC, 0400, 0420, 0430, 0500 UTC
5.855	RTTY		USSR EMBASSY IN WASHINGTON, D.C.
5.865	VOICE		SPY STATION, SPANISH FEMALE VOICE, SENDS 5-DIGIT TRAFFIC, 0400, 0500 UTC (WILL REPEAT TRAFFIC SENT ON 6.825, 9.255 AT 0200, 0300 UTC
	SSB	VJN	ROYAL FLYING DOCTORS, AUSTRALIA, 0800-1700 UTC
5.867	RTTY	YIL68	IRAQ NEWS AGENCY
5.870	CW	NMN	COMMSTA, NORFOLK, VA (Sends a lot of 5-digit coded traffic) CQ CQ CQ DE NMN/NAM/NAR WEATHER QRV
5.882	VOICE		SPY STATION, SPANISH FEMALE VOICE, SENDS 5-DIGIT TRAFFIC, 0400 UTC
	CW		SPY STATION, SPANISH FEMALE VOICE, SENDS 5-DIGIT TRAFFIC, 0400, 0500 UTC

Frequency	Mode	Call Sign	Service / Times
5.895 MHz	ARQ		INTERPOL FREQUENCIES
	SSB	VJC	ROYAL FLYING DOCTOR, AUSTRALIA, 0700-1900 UTC
5.896			SPY STATION, SPANISH FEMALE VOICE, SENDS 5-DIGIT TRAFFIC, 0300, 0330, 0430, 0500, 0600 UTC
5.905	AM		RADIO MOSCOW, (West Coast), 0630-0900 UTC (November thru February)
5.9165	CW	NMN	CQ CQ CQ DE NMN/NAM/NRK/NAR/GXS/AOK QRU
5.922	CW	X	HF BEACON LOCATED IN PRAGUE
5.930 MHz	AM		RADIO PRAGUE, 0300-0500 UTC
	AM		VOICE OF FREE CHINA, 2200-2300 UTC
	AM		RADIO CZECHOSLOVAKIA, (English Service), 1800-1827, 2100-2130, 0100-0130 UTC, (Spanish Service), 2300-2327, 0200-0227 UTC
	VOICE		SPY STATION, SPANISH FEMALE VOICE, SENDS 4-DIGIT TRAFFIC, 0000, 0200, 0330, 0400, 0430, 0500, 2300 UTC (WILL BROADCAST ON 10.665 AND 11. 532 AT SAME TIME)
5.935	AM		RADIO LATVIA, 0300-0500 UTC
5.9385	CW	BFL	RVD RVD RVD DE BFL BFL BFL QSA ? ? ? QTC ? QSA NIL K
5.945	AM		RADIO AUSTRIA INTERNATIONAL, 1830-1900, 2130-2200 UTC
5.950	AM		VOICE OF AMERICA, (china service) 0500 UTC
	AM		VOICE OF FREE CHINA, TAIWAN, (English Service), 0200-0400, 0700-0800 UTC

Frequency	Mode	Call Sign	Service / Times
5.950 MHz	AM		VOICE OF GUYANA, 0730-0200 UTC., (Operates from 04:30 AM to Midnight, local time. Except on Fridays when sign off is at 01:00 AM Saturday)
	AM	WYFR	FAMILY RADIO NETWORK, 1000-1400 UTC
5.955	AM		BRITISH BROADCASTING CORPORATION, 1300 UTC
	AM		RADIO FREE EUROPE (Romanian), 0400-0600, 1800-2200 UTC
	AM		VOICE OF AMERICA, (Mandarin to East Asia), 1100-1600 UTC, (Polish to Europe), 0530-0630 UTC
5.960	AM		RADIO JAPAN, 0100-0200 UTC
	AM		RADIO DEUTSCHE WELLE, (Radio Germany), 0500-0600 UTC
	AM		RADIO CANADA INTERNATIONAL, 2330-0030, 2200-2230, (Saturday & Sunday) 2330-0100 UTC
5.965	AM		VOICE OF AMERICA, (English to Middle East/Europe service), 0200-0500 UTC
	AM		BRITISH BROADCASTING CORPORATION, 0000-0300, 1100-1200 UTC
	AM		VOICE OF AMERICA, (English to VOA Europe), 0300-0330 UTC
	AM		RADIO CUBA 0000-0800 UTC, (Side Band Service), (Beam to U.S.)
5.970	AM		RADIO FREE EUROPE (Romanian), 0500-1500 UTC
5.975	AM		BRITISH BROADCASTING CORPORATION 0000-0730, 1630-1830 UTC
	AM		RADIO SOUTH KOREA, (KBS), 1600-1700 UTC

Frequency	Mode	Call Sign	Service / Times
5.9775MHz	AM		THE VOICE OF LEBANON, (English Service) 0900-1000, 1315-1330, 1815-1830 UTC
5.985	AM		VOICE OF AMERICA, (English to Pacific service) 1000-1200 UTC
	AM		RADIO FREE EUROPE, (Romanian), 0400-0700, (Hungarian) 600-2200 UTC
	AM	WYFR	FAMILY RADIO NETWORK, 0000-0100, 0500-0700 UTC, (Spanish Service), 0100-0500 UTC
5.990	AM		RADIO ROMANIA INTERNATIONAL, 2130-2300 UTC
5.995	AM		VOICE OF AMERICA, (English to American Republics service) 0000-0230 UTC, (Service to Europe), 0330-0700 UTC
	AM		RADIO AUSTRALIA, 1130-2100 UTC

DATE	FREQ.	MODE	TIME	STATION	SIGNAL	COMMENTS	SWL SENT	SWL REC'D

Frequency	Mode	Call Sign	Service / Times
6.000 MHz	AM		RADIO MOSCOW, (East Coast) 0000-0500 UTC, (November thru February)
6.005	AM		VOICE OF AMERICA, (South Africa beaming)
	AM		BRITISH BROADCASTING CORPORATION, 0000-0330, 0500-0545 UTC
	AM	CIQX	CIQC SHORTWAVE OF CANADA, 0000-2359 UTC
6.010	AM		RADIO MEXICO, 0200-0700 UTC
	AM		VOICE OF AMERICA, (Urdu to South Asia), 0100-0130 UTC, (Hindi to South Asia) 0030-0100 UTC
	AM		BRITISH BROADCASTING CORP., 0200-0300 UTC
6.012	AM		OPERATION DEEP FREEZE, (U.S. Navy, Start-up time around February 1992), (No Broadcasts Schedules Available at this Time),(Per Commander M.R. Reed, U.S. Navy)
6.015	AM		VOICE OF AMERICA, (Georgian to the USSR), 0315-0330 UTC
	AM		RADIO AUSTRIA INTERNATIONAL, 0530- 0600, 0630-0700 UTC
6.020	AM		RADIO NETHERLANDS INTERNATIONAL, (English Service), 0030-0125 UTC, (Spanish Service), 1130-1155, 2330-0025 UTC
	AM		VOICE OF AMERICA, 0000-0600 UTC

Frequency	Mode	Call Sign	Service / Times
6.020 MHz	AM		VOICE OF AMERICA, (French to Africa), 0530-0700 UTC (Hausa to Africa), 0500-0530 UTC, (Portuguese to Africa), 0430-0500 UTC
6.025	AM		RADIO FREE EUROPE, 1100-1200 UTC
6.030	AM		VOICE OF AMERICA, (Spanish to Cuba) 0600-0930 UTC, (Spanish to American Republics) 0100-0400 UTC (Indonesian to East Asia), 1330-1400, (China) 1500-1600 UTC
6.035	AM		THE VOICE OF IRAN, (English Service) VOICE OF AMERICA, (English to Africa ser.) 0300-0700 UTC
6.040	AM		VOICE OF AMERICA, (English to Middle East/Europe service), 1700-2200 UTC (Spanish to American Republics) 0930-1130 UTC, (Russian to USSR) 0300-0400 UTC
6.045	AM		BRITISH BROADCASTING COPR., 0900-1500 UTC
	AM		RADIO MOSCOW, (East Coast), 0000-0500 UTC, (November thru February)
	AM		RADIO DEUTSCHE WELLE, (Radio Germany), 0300-0400, 0500-0600 UTC
6.050	AM		RADIO FREE EUROPE, (Romanian), 0000-0200, 2300-2400UTC

Frequency	Mode	Call Sign	Service / Times
6.050 MHz	AM	HCJB	THE VOICE OF THE ANDES (Spanish Service) 1030-0500
6.055	AM		THE VOICE OF BAGHDAD, (English Service), 1700-1900 UTC
	AM		RADIO DEUTSCHE WELLE, (Radio Germany), 0300-0400 UTC
	AM		RADIO CZECHOSLOVAKIA, (English Service), 1800-1827, 1930-1957, 2100-2130, 2200-2225 UTC
	AM		RADIO EXTERIOR DE ESPANA, (Spanish Service) 2300-0500 UTC
6.060	AM		RADIO AUSTRALIA, 1600-2130 UTC
	AM		VOICE OF AMERICA, (English to Middle East/Europe service), 0500-0700, (Latvian to Europe), 1630-1700 UTC, (Estonian to Europe), 1600-1630 UTC, (Russian to USSR) 2200-2400 UTC
6.065	AM	WYFR	FAMILY RADIO NETWORK, 0100-0500 UTC
6.070	AM		RADIO FREE EUROPE, (Bulgarian), 1700-2100 UTC
6.075	AM		VOICE OF AMERICA , (Spanish to Cuba) 0930-1200 UTC
	AM		RADIO BERLIN, GERMANY, 0600 UTC
6.080	AM		RADIO AUSTRALIA, 1100-2100 UTC
	AM		VOICE OF AMERICA, (Serbo-Croatian to Europe), 0445-0500 UTC

Frequency	Mode	Call Sign	Service / Times
6.080 MHz	AM		VOICE OF AMERICA, (Georgian to the USSR), 0315-0330 UTC, (Hungarian to Europe) 0530-0600 UTC
	AM	HCJB	THE VOICE OF THE ANDES (Spanish Service) 0200-0500
6.085	AM		VOICE OF AMERICA, (Russian to USSR) 2100-2300 UTC
	AM		RADIO DEUTSCHE WELLE, (Radio Germany), 0300-0400 UTC
	AM	WYFR	FAMILY RADIO SERVICE, (Spanish Service), 1000-0100 UTC
6.090	AM		VOICE OF AMERICA, (Ukrainian to the USSR), 0300-0500 UTC
6.095	AM		VOICE OF AMERICA, (English to Middle East/Europe service) 0100-0200, 0600-0700 UTC (Russian to USSR) 2000-2200 UTC
	AM		POLISH RADIO WARSAW, (English Service) 1500-1525, 1730-1755 UTC (Spanish Service) 1430-1455, 2130-2155 UTC
	AM	KNLS	THE NEW LIFE STATION 0800-0900 UTC
6.105	AM		RADIO FREE EUROPE, (Romanian), 0000-2400 UTC
	AM	WYFR	FAMILY RADIO NETWORK, (Spanish service), 0800-1100 UTC
6.110	AM		VOICE OF AMERICA, (English to Pacific serv.) 1100-1500 UTC

Frequency	Mode	Call Sign	Service / Times
6.110 MHz	AM		(Korean to East Asia), 2130-2200 UTC
	AM		BRITISH BROADCASTING CORP., WORLD SERVICE, (Spanish Service) 0000-0200, 0300-0430 UTC
6.115	AM		RADIO FREE EUROPE, (Romanian), 2000-2400 UTC
	AM		RADIO BERLIN, GERMANY, 0600 UTC
6.120	AM		RADIO JAPAN, 1100-1200 UTC
	AM		RADIO DEUTSCHE WELLE, (Radio Germany), 0300-0400, 0500-0600 UTC
6.125	AM		VOICE OF AMERICA, (Service to Europe) 0300-0400 UTC, (English to Africa serv.) 0600-0700 UTC
	AM		SWISS RADIO INTERNATIONAL, 0200-0230 UTC
	AM		RADIO EXTERIOR DE ESPANA (Spanish Service) 0900-1900 UTC
6.130	AM		VOICE OF AMERICA, (English to Caribbean), 0000-0200 UTC (Serv. to Latin America), 0200-0300UTC
	AM		RADIO DEUTSCHE WELLE, (Radio Germany), 0500-0600 UTC
6.135	AM		RADIO FREE EUROPE, (Kirghiz), 0000-0100 UTC, (Romanian) 0200-0400 UTC, (Kirghiz), 2300-2400 UTC

Frequency	Mode	Call Sign	Service / Times
6.135 MHz	AM		SWISS RADIO INTERNATIONAL, (English Service), 0000-0030, 0200-0230, 0400-0430 UTC, (Spanish Service), 0030-0100, 0230-0300 UTC
	AM		RADIO SANTA CRUZ, BOLIVIA, 1000-1200 UTC
	AM		RED CROSS BROADCASTING SERVICE, (English), 0310-0327 UTC
	AM		RADIO SOUTH KOREA, (KBS), (English Service), 1930-2000 UTC
	AM		POLISH RADIO WARSAW, (English Service) 1400-1425, 1600-1625, 1730-1755, 1900-1925 UTC (Spanish Service) 1430-1455, 2130-2155 UTC
6.140	AM		VOICE OF AMERICA, (English to MIddle East/Europe service) 0400-0700 (Georgian to USSR), 2000-2030 UTC (Russian to USSR), 1800-2000, 2100-2400 UTC
	AM		ISLAMIC REPUBLIC OF IRAN BROADCASTING, (English Service) (Beaming Europe), 1930-2030 UTC, (Spanish Service), 2030-2130 UTC
6.145	AM		VOICE OF AMERICA, (English to Africa service), 0300-0430 UTC
	AM		RADIO DEUTSCHE WELLE, 0300-0500 UTC
6.150	AM		RADIO CANADA INTERNATIONAL 0515-0600 UTC, (English Service)
	AM		VATICAN RADIO, VATICAN CITY, 0030-0130 UTC

Frequency	Mode	Call Sign	Service / Times
6.150 MHz	AM		VOICE OF AMERICA, (Ukrainian to the USSR), 1600-1800 UTC (Service to USSR), 0300-0500 UTC, (Service to Europe), 1900-2100 UTC
6.155	AM		RADIO AUSTRIA INTERNATIONAL, 0530-0600, 0830-0900, 1130-1200 1530-1600, 1830-1900, 2130-2200 UTC
6.160	AM		VOICE OF AMERICA, (Russian to USSR) 2200-2400 UTC (Farsi to Middle East), 1700-1900 UTC (Turkish to Middle East), 2000-2100 (Ukrainian to USSR), 0300-0500 UTC
6.165	AM		RADIO NETHERLANDS INTERNATIONAL, (English Service), 0030- 0125 UTC, (Spanish Service), 0230-0325, 0430-0525 UTC
	AM		SWISS RADIO INTERNATIONAL, 0730-0800 UTC
6.170	AM		RADIO FREE EUROPE, (Bulgarian), 1600-1800 UTC
6.175	AM		BRITISH BROADCASTING CORPORATION, 2300-0330 UTC
6.180	AM		VOICE OF AMERICA, (English to Middle East/Europe service), 1630-1700 UTC, (Hausa to Africa), 0500-0530 UTC (Russian to USSR) 0300-0400 UTC (French to Africa), 0530-0700 UTC
	AM		BRITISH BROADCASTING CORPORATION, 0300-1530, 1700-2200 UTC

Frequency	Mode	Call Sign	Service / Times
6.190 MHz	AM		BRITISH BROADCASTING CORPORATION, 0500-1800 UTC
	AM		VOICE OF AMERICA, (Spanish to the American Republics), 0100- 0300 UTC
6.195	AM		BRITISH BROADCASTING CORPORATION, 0300-0630, 1500-2200 UTC
6.203	USB		KOREAN FISHING BOATS
6.205	AM		RADIO MOSCOW, (English Service), 0230 UTC
6.2155	SSB		MARITIME SAFETY AND DISTRESS
6.218	USB		INTERNATIONAL SSB RADIO TELEPHONY, (Receive on 6519.00)
6.2216	USB		SHIP TO SHIP, SHIP TO SHORE COMMUNICATIONS
6.225	VOICE		SPY STATION, SPANISH FEMALE VOICE, SENDS 5-DIGIT TRAFFIC, 0700 UTC
6.263	USB		JAPANESE MARITIME AGENCY
6.276	CW		ELL CGH DE ELL58 QSV ? QRU ?
6.2995	USB/ARQ		COMSAC, U.S. COAST GUARD
6.305	AM		POLISH RADIO WARSAW, (English Service) 1300-1355 UTC
6.315	CW		WNU DE WNU SELCAL 1109 6315/6263R5
6.316	ARQ		NMN
6.317	ARQ		PCH
	ARQ	WLO	(Receive on 06265.50)
6.318	ARQ	KLB	
6.3185	ARQ	SOJ	
6.319	ARQ	WLO	(Receive on 06267.50)
6.320	ARQ	KPH	
6.321	ARQ	WLO	
	ARQ	FFT31	
6.321	ARQ	WLO	(Receive on 06270.00)
6.323	ARQ	WLO	(Receive on 06272.00)
6.324	ARQ	WCC	
	ARQ	NMC	
6.325	ARQ	KPH	
	ARQ	WLO	(Receive on 06274.50)
6.3266	CW	WNU32	CQ CQ CQ DE WNU32 WNU32 WNU32 QSX 4 6 8 12 MHZ OBS ?

Frequency	Mode	Call Sign	Service / Times
6.330 MHz	FAX	CFN	CFN HALIFAX, WILL DURING THE HOUR STOP SENDING FAX PICTURES AND CHANGE OVER TO RTTY FOR SENDINGMESSAGES AND UP-TO-DATE WEATHER REPORTS. ALSO TRANSMITS ON 10536 MHZ.
	RTTY	CFN	75 BAUD BAUDOT RXREV ON, SENDS WEATHER REPORTS AND TRAFFIC FAX CFH HALIFAX, NS, WEATHER
6.341	ARQ	WLO	
6.3445	CW	WLO	DE WLO 2 OBS ? AMVERS ? QSX 4 6 8 12 16 22 25R172 MHZ NW ANS C3/4 K
6.348	CW	HWN	VVV DE HWN
63505	CW	VID	WEATHER/AMVER
6.351	CW	P2M	WEATHER/AMVER
6.3515	CW	WMH	WEATHER/AMVER
	CW	VIB	WEATHER/AMVER
6.3588	ARQ	DHS	
6.365	CW	KFS	CQ DE KFS KFS KFS/B QSX 8 12 16 22 MHZ K
6.369	CW	D3E41	CQ DE D3E41/51/52 QSX 6 AND 8 MHZ C. 6 ON RTF 542 821 1621 AND 2207 AR K
	CW	KLC	WEATHER/AMVER, (Galveston, Texas)
6.370	CW	UBE	WEATHER/AMVER
6.371	CW	GYU	DE GYU QSX 6 8 12
6.3764	CW	WCC	VVV VVV DE WCC WCC BT OBS ? QSX 6 8 12 MHZ K
6.3825	CW	EAD2	DE EAD2/EAD3 QSX 4/8 MHZ CG AR K
6.3838	CW	NMC	CQ CQ CQ DE NMC NMC NMC QRU ? AMVER ? OBS ? GOVT TFC ? PLEAD ? DE NMC NMC QRU ? K
6.3854	CW	CKN	NAWS DE CKN II ZKR F1 2386 4158 6245 8315 12383 16564 22191 25124 KHZ AR
6.3857	CW	PPR	VVV DE PPR PPR PPR PSE QSX 4 MHZ K
6.3877	CW	HKC	CQ CQ CQ DE HKC HKC HKC QRU ? QSX 8/12/16 MHZ ON CHANNELS 5/6 K
	CW	ZSJ	CQ CQ CQ DE ZSJ ZSJ ZSJ QSX AMVER ANS 4/6/8/12 AND 22533 KHZ CH 3/4/9/10 ARK
6.3905	CW	WNU42	CQ CQ CQ DE WNU42 WNU42 WNU42 QSX 4 6 8 12 MHZ OBS ?
6.3943	CW	ZLB	VVV DE ZLB3 ZLB3 QSX

Frequency	Mode	Call Sign	Service / Times
6.400 MHz	VOICE		SPY STATION, SPANISH FEMALE VOICE, SENDS 5-DIGIT TRAFFIC, 1500 UTC
6.407	CW	VIT	WEATHER/AMVER
6.4115	CW	KLB	CQ CQ DE KLB KLB QSX 4 6 8 12 16 AND 22 MHZ OBS ? SITOR SELCAL 1113 AR K
6.4155	CW	7TF4	CQ CQ CQ DE 7TF4 7TF4 7TF4 QSX 6 MHZ C. 5/6/8 TKS K
6.416		FEC	WEATHER REPORTS
	CW	WLO	DE WLO END TFC LIST NW ANS MF/HF K
6.4222	CW	FFL	V CQ CQ DE FFL2/3/4 FFL2/3/4 QSX 4 AND 8 MHZ CHANNEL 1 TO 6 K
6.425	AM		THE VOICE OF LEBANON (English Service) 0900-1000, 1315-1330, 1815-1830 UTC
6.428	CW	VIX	WEATHER/AMVER
6.4305	CW	CFH	VVV VVV VVV DE CFH CFH CFH C13L C13L C13L
6.43635	FAX	GYA	NORTHWOOD, UNITED KINGDOM, 0000-2400 UTC
6.441	VOICE		SPY STATION, SPANISH FEMALE VOICE, SENDS 5-DIGIT TRAFFIC, 0530 UTC
6.4428	CW	XFS2	XCCE XCCP XCEP XCNP XCRN XCRV DE XSF2 QTC QSX 4206 MHZ K
6.4466	CW	WLO	DE WLO 2 OBS ? AMVERS ? QSX 4 6 8 12 16 22 25R172 MHZ NW ANS C3/4 K
6.4477	CW	VAS	CQ CQ CQ DE VAS VAS VAS
6.453	FAX	NPM	HONOLULU, HAWAII
6.454	VOICE		SPY STATION, SPANISH FEMALE VOICE, SENDS 5-DIGIT TRAFFIC, 0530 UTC SENDS 4-DIGIT TRAFFIC, 0500 UTC
6.4558	CW	UDK	4LS 4LS DE UDK2 UDK2 QSX 4189 K
6.456	CW	CKN	VVV VVV VVV DE CKN CKN CKN C13E C13E
6.4583	CW	JOR	CQ CQ CQ DE JOR JOR JOR QSX 6 MHZ K
6.460	CW	LSA	VVV DE LSA PSE QSX BT
6.462	CW	FUM	VVV DE FUM
6.4635	CW	HKB	CQ CQ CQ DE HKB HKB HKB QSX ON 6.273/8.364/12.546 MHZ K
6.464	RTTY		75 BAUD, 6-BIT RTTY, RXREV OFF
	CW	VIA	WEATHER/AMVER
6.4688	CW	JCS	CQ CQ CQ DE JCS JCS JCS QSX 6 MHZ

Frequency	Mode	Call Sign	Service / Times
6.476 MHz	CW	KPH	VVV DE KPH QSX 4 6 8 12 16 22 MHZ DE KPH QSX 4 6 8 12 16 22 K
6.479	ARQ	WLO	(Receive on 06259.00)
6.480	AM		RADIO SOUTH KOREA, (KBS), (English Service), 2030-2130 UTC (Spanish Service), 2215-2300 UTC
6.4845	CW	WSC	CQ CQ DE WSC WSC QSX 6 8 12 MHZ OBS ? DE WSC K
6.4923	CW	VCS	VVV VVV VVV CQ DE VCS VCS VCS QSX 4 6 AND 8 MHZ CHNL 11/5/61 (Duel on 08438 MHZ)
6.4938	CW	VAI	CQ CQ CQ DE VAI VAI VAI QSX 4/6/8/12/16 MHZ CH4/5 OBS/AMVER/QRJ/WESTREG ? VAI SITOR SELCALL 1.00581 QRU ? K
6.494	CW	NMN	NEWS NET
6.4958	CW	WNU32	CQ CQ CQ DE WNU32 WNU32 WNU32 QSX 4 6 8 12 16 MHZ OBS ?
6.4955	ARQ	KFS	
6.496	ARQ	NMO	
6.4965	ARQ	PCH	
6.497	ARQ	VIP	
6.4975	ARQ	VCS	
	ARQ	JNA	
6.498	ARQ	WCC	
6.4985	ARQ	KLC	
6.500	ARQ	NMN	
6.5005	ARQ	WCC	
6.501	ARQ	VIS63	
	USB		COMSAC ALASKA, U.S. COAST GUARD
	USB		NATIONAL OCEANIC AND ATMOSPHERIC ADMINISTRATION, (NOAA), (Uses Computer Voice)
6.5025	ARQ	KLC	
6.503	ARQ	WNU	
6.504	ARQ	NMC	
6.505	ARQ	NMN	
6.506	ARQ	KLC	
	ARQ	WLO	
6.520	USB		SKYKING NET
6.5219	USB		SHIP TO SHIP, SHIP TO SHORE COMMUNICATIONS.

Frequency	Mode	Call Sign	Service / Times
6.549 MHz	AM		THE VOICE OF LEBANON, (English Service), 0900-1000, 1315-1330, 1815-1830 UTC
6.577	USB		INTERNATIONAL AIRLINES
6.586	USB		INTERNATIONAL AIRLINES
6.604	USB		INTERNATIONAL AIRLINES WEATHER
6.6285	USB		AIR LINE TRAFFIC
6.673	USB		NOAA HURRICANE HUNTERS
6.678	USB		AERONAUTICAL WEATHER REPORTING
6.6798	USB		WX VOICE, HONOLULU, HAWAII
6.680	USB		SKYKING BROADCASTS
6.6866	USB		JAPAN AIRLINES
6.6962	USB		SKYKING BROADCASTS
6.7115	USB		USAF MIDDLE EAST
6.7135	USB		US PILOT'S AIR FORCE
6.715	USB	K83	U.S. NAVY, (C17, PAPA RESCUE, ETC.)
6.712	USB		SKYKING BROADCASTS
6.730	USB		SKYKING BROADCASTS
	USB		LOOKING GLASS
6.735	CW	X	HF BEACON LOCATED IN PRAGUE
6.7375	RTY	ETD3	RY'S, 50 BAUD, BAUDOT
6.738	USB		USAF MIDDLE EAST
	USB		MILITARY AIRLIFT COMMAND
	USB		US PILOT'S AIR FORCE
6.745	VOICE		SPY STATION, SPANISH FEMALE VOICE, SENDS 5-DIGIT TRAFFIC, 0240 UTC
6.748	USB/CW		RUSSIAN AIRLINE AEROFLOT
6.750	USB		SKYKING BROADCASTS
6.7525	USB		SKYKING BROADCASTS
6.753	USB		USAF MIDDLE EAST
	USB		US PILOT'S AIRFORCE
6.757	USB		LOOKING GLASS
	USB		SKYKING BROADCASTS
6.760	USB		SKYKING BROADCASTS
6.761	USB		LOOKING GLASS
	USB		SKYKING BROADCASTS
6.762	USB		SKYKING BROADCASTS
6.76742	USB	D1T13	U.S. MILITARY, (0500 UTC) (Call signs change daily)
6.7677	CW	P7X	QRA DE P7X, (Station sends coded traffic, CW, RTTY)

Frequency	Mode	Call Sign	Service / Times
6.7785 MHz	PAKT		MARS PACKET NET
6.785	USB		U.S. ARMY, MILITARY (Call signs change daily)
	VOICE		SPY STATION, ENGLISH/SPANISH VOICE, SENDS 5-DIGIT TRAFFIC, 0400, 0600, 2200 UTC (IF 0300 SKED ON 5.896 IS KEPT, THE 0400 SKED IS ACTIVE)
	CW		SPY STATION, MORSE CODE, SENDS 5-DIGIT TRAFFIC, 0200, 0230, 0300, 0400, 0500, 0530, 0600, 0800, 0830, 1055, 1100 UTC
	SSB	VJN	ROYAL FLYING DOCTOR, AUSTRALIA, 0800-1700 UTC
	LSB	WUK410	WUK410
6.787	VOICE		SPY STATION, SPANISH FEMALE VOICE, SENDS 3-2 DIGIT TRAFFIC, 0000, 0300, 0530 UTC
6.789	VOICE		SPY STATION, SPANISH FEMALE VOICE, SENDS 5-DIGIT TRAFFIC, 0000, 0500 UTC 4-DIGIT TRAFFIC 2300 UTC
6.790	FAX	YMA	WEATHER, ANKARA, YUGOSLAVIA
6.792	ARQ		INTERPOL FREQUENCIES
6.794	VOICE		SPY STATION, SPANISH FEMALE VOICE, SENDS 5-DIGIT TRAFFIC, 0500 UTC
6.797	VOICE		SPY STATION, SPANISH FEMALE VOICE, SENDS 5-DIGIT TRAFFIC, 0000, 0200, 0300, 0400, 0500 UTC 4-DIGIT TRAFFIC, 0530 UTC
6.803	VOICE		SPY STATION, SPANISH FEMALE VOICE, SENDS 4-DIGIT TRAFFIC, 0200, 0230, 0300, 0400, 0500, 2200, 2300 UTC 5-DIGIT TRAFFIC, 0100 UTC
6.805	VOICE		SPY STATION, SPANISH FEMALE VOICE, SENDS 5-DIGIT TRAFFIC, 0020, 0200, 2100 UTC
6.816	VOICE		SPY STATION, SPANISH FEMALE VOICE, SENDS 5-DIGIT TRAFFIC, 0600 UTC
6.822	CW	EAD	DE EAD2/EAD4 QSX 6 AND 8 MHZ CG AR K

Frequency	Mode	Call Sign	Service / Times
6.825 MHz	RTTY		U.S. ARMY MARS
	VOICE		SPY STATION, GERMAN FEMALE VOICE, SENDS 5-DIGIT TRAFFIC, 0500, 0530 UTC SPANISH FEMALE VOICE, 0100, 0300 0500, 0530, 0540, 0600, 0630, 1000 UTC (ALSO BROADCAST ON 10.245 AT SAME TIME)
	CW		SPY STATION, MORSE CODE, SENDS 5-DIGIT TRAFFIC, 0930 UTC
	SSB	VJO	ROYAL FLYING DOCTOR, AUSTRALIA, 0700–1700 UTC
6.826	USB		SKYKING BROADCASTS
	VOICE		SPY STATION, SPANISH FEMALE VOICE, SENDS 3-2 DIGIT TRAFFIC, 0600 UTC 4-DIGIT TRAFFIC, 0100, 0200, 0230, 0400, 0530, 0630 UTC 5-DIGIT TRAFFIC, 0130, 0200, 0500, 0600, 0630, 0900 UTC
6.830	VOICE		SPY STATION, SPANISH FEMALE VOICE, SENDS 5-DIGIT TRAFFIC, 0530 UTC
6.8376	AM		VOICE OF AMERICA (Relay link)
6.840	USB		SKYKING BROADCASTS
	VOICE		SPY STATION, SPANISH FEMALE VOICE, SENDS 4-DIGIT TRAFFIC, 0000, 0130, 0200, 0230, 0400, 0500 2200, 2300 UTC (BROADCAST ON 9.958 AT SAME TIME)
	SSB	VJY	ROYAL FLYING DOCTOR, AUSTRALIA, 0800-2400 UTC
6.845	VOICE		SPY STATION, SPANISH FEMALE VOICE, SENDS 5-DIGIT TRAFFIC, 0200, 0230, 0300, 0330, 0400, 0430 UTC
	SSB	VJJ	ROYAL FLYING DOCTOR, AUSTRALIA, 0800-1700 UTC
6.8475	AM		VOICE OF AMERICA (Relay link)
6.850	FAX	WLO	MOBILE, AL, WEATHER
		CW	SPY STATION, MORSE CODE, SENDS 5-DIGIT TRAFFIC, 0500, 0700 UTC
6.855	VOICE		SPY STATION, SPANISH FEMALE VOICE, SENDS 5-DIGIT TRAFFIC, 0700 UTC
6.863	USB		SKYKING BROADCASTS

Frequency	Mode	Call Sign	Service / Times
6.866 MHz	SSB	VJN	ROYAL FLYING DOCTOR, AUSTRALIA, 0800-1700 UTC
6.872	FAX	LRB79	B. AIRES, ARG, PRESS
6.873	USB/LSB		VOICE OF AMERICA, 0400 UTC (Relay link)
6.878	RTTY		75 BAUD, BAUDOT, RXREV ON
6.880	SSB	VKJ	ROYAL FLYING DOCTOR, AUSTRALIA, 0600-1700 UTC
6.887	VOICE		SPY STATION, SPANISH FEMALE VOICE, SENDS 4-DIGIT TRAFFIC, 0500 UTC SENDS 5-DIGIT TRAFFIC, 0500, 0600 UTC
6.890	VOICE		SPY STATION, SPANISH FEMALE VOICE, SENDS 5-DIGIT TRAFFIC, 0600, 0700 UTC
	SSB	VNZ	ROYAL FLYING DOCTOR, AUSTRALIA, 0800-1700 UTC
6.896	VOICE		SPY STATION, SPANISH FEMALE VOICE, SENDS 5-DIGIT TRAFFIC, 0600, 0630, 0700 UTC
6.900	RTTY		75 BAUD, BAUDOT, RXREV OFF, (Weather Reports)
6.901	FAX	AOK	WEATHER, ROTA, SPAIN
6.9025	FAX	CLN49	WEATHER, HAVANA, CUBA
6.905	ARQ		INTERPOL FREQUENCIES
6.910	AM		RADIO DUBLIN LIMITED, English Broadcasts 24 hours daily.
6.920	SSB	VJC	ROYAL FLYING DOCTOR, AUSTRALIA, 0700-1900 UTC
6.9245	ARQ		DEPT. OF STATE IN WASHINGTON, D.C.
6.925	CW	KKN50	QRA QRA QRA DE KKN50 KKN50 KKN50 QSX 6/10/12/16 K KRC81 KRC81 KRC81 DE KKN50 KKN50 KKN50 QSX 6/10/12/16 K (Will transmit call once every minute)
	SSB	VJB	ROYAL FLYING DOCTOR, AUSTRALIA, 0700-1700 UTC
6.933	VOICE		SPY STATION, SPANISH FEMALE VOICE, SENDS 4-DIGIT TRAFFIC, 0300, 0400, 0500 UTC (WILL BROADCAST ON 10.665 AT SAME TIME)

Frequency	Mode	Call Sign	Service / Times
6.935 MHz	VOICE		SPY STATION, SPANISH FEMALE VOICE, SENDS 4-DIGIT TRAFFIC, 0000, 0400, 0500, 0600 UTC ENGLISH FEMALE VOICE, SENDS 3-2 DIGIT TRAFFIC, 0030, 0100 UTC 5-DIGIT TRAFFIC 0400, 0530 UTC
6.942	VOICE		SPY STATION, SPANISH FEMALE VOICE, SENDS 5-DIGIT TRAFFIC, 0600 UTC
6.944	FAX	CKN	VANCOUVER, BC, WEATHER
6.945	SSB	VJJ	ROYAL FLYING DOCTOR, AUSTRALIA, 0800-1700 UTC
6.950	SSB	VJD	ROYAL FLYING DOCTOR, AUSTRALIA, 0700-1700 UTC
6.960	SSB	VKL	ROYAL FLYING DOCTOR, AUSTRALIA, 0700-1700 UTC
6.964	VOICE		SPY STATION, ENGLISH FEMALE VOICE, SENDS 5-DIGIT TRAFFIC, 0200, 0240, 0300 UTC
6.965	SSB	VJI	ROYAL FLYING DOCTOR, AUSTRALIA, 0800-1700 UTC
6.9665		TDM	75 BAUD, TDM 2:8 RXREV OFF
6.970	USB	NAU	USN, CEIBA, PTR
6.980	ASCII		110 BAUD, ASCII, RXREV ON
	CW	ZKLT	3XZY DE ZKLT QTC ? QRU ? QSV KK
6.988	USB	AAH	HF/MARS RADIO STATION, FT LEWIS, WA
6.985	USB		ARMY MARS
	USB	D1T23	U.S. MILITARY, (0500 UTC)
6.98950	LSB		U.S. ARMY MARS, 2200 UTC
6.992	VOICE		SPY STATION, ENGISH FEMALE VOICE, SENDS 3-2 DIGIT TRAFFIC, 0530 UTC 5-DIGIT TRAFFIC, 0500 UTC
6.99550	USB	AIR	HF/MARS 2045TH COMMUNICATIONS GROUP, ANDREWS AFB, WASHINGTON, DC
6.997.50	USB	WAR	HF/MARS RADIO STATION, FT DETRICK, MD

Frequency	Mode	Call Sign	Service / Times
7.000 MHz	**SSB/CW**		**START OF AMATEUR RADIO 40 METER BAND (Ends 7.300 MHz)**
7.035	CW		QUARTER CENTURY WIRELESS ASSOCIATION, 2000 HOURS, WEDNESDAY
7.047	CW	W1AW	CW BULLETINS
7.075	PACKET		AMATEUR RADIO PACKET NETS
7.095	RTTY		ARRL RTTY BULLETINS
	AM		VOICE OF REBELLIOUS IRAQ, 0430-0730, 1730-2030 UTC
7.100	VOICE		SPY STATION, ENGLISH MALE VOICE, SENDS 3-2 DIGIT TRAFFIC, 0400 UTC
7.105	AM		VOICE OF AMERICA, (Russian to USSR), 0300-0400 UTC
	AM		BRITISH BROADCASTING CORPORATION, 0400-0600 UTC
	AM		VOICE OF AMERICA, (Urdu to South Asia), 0100-0200 UTC
	AM		UNITED NATIONS RADIO, (Sunday Service Only), 0830-0900 UTC
7.110	AM		VOICE OF AMERICA, (Service to Middle East), 2300-2330 UTC
7.11480	LSB	"JUG"	DRUG TRAFFIC (Different call signs are commonly used)
7.115	AM		RADIO FREE EUROPE, (Hungarian), 1600-2200 UTC
	AM		VOICE OF AMERICA, 0100-0300 UTC
	AM		RADIO MOSCOW, (East Coast), 0000-0400 UTC, (November thru February)
7.120	AM		BRITISH BROADCASTING CORPORATION, 1300 UTC
	AM		VOICE OF AMERICA, (English to Pacific service)

Frequency	Mode	Call Sign	Service / Times
			2200-0100 UTC
7.120 MHz	VOICE		SPY STATION, SPANISH FEMALE VOICE, SENDS 4-DIGIT TRAFFIC, 0430 UTC
7.125	AM		VOICE OF AMERICA, (English to Pacific service) 1400-1500 UTC (English to VOA Europe), 0100-1530, 1600 1800 UTC (Russian to USSR), 2300-2400 UTC
	AM		UNITED NATIONS RADIO, (Sunday Service Only), 0830-0900 UTC
7.130	AM		VOICE OF AMERICA, (Service to USSR), 2200-2300 UTC (Service to Europe), 0430-0530 UTC
7.135	AM		BRITISH BROADCASTING CORPORATION, 0030-0330 UTC
7.140	AM		RADIO AUSTRALIA, 0900-1000, 1100-1200 UTC
	AM		RADIO JAPAN, (Asia), 1700-1800 UTC, (Africa), 1900-1930,
7.145	AM		RADIO FREE EUROPE (Kazak) 0000-0200, 2300-2400 UTC
	AM		BRITISH BROADCASTING COPRORATION, 2300-0030 UTC
	AM		POLISH RADIO WARSAW, (English Service) 1300-1355, 1500-1525, 1900-1925 UTC (Spanish Service) 1430-1455, 2130-2155 UTC
7.150	AM		BRITISH BROADCASTING CORPORATION, 0600-0830, 1300 UTC
	AM		RADIO DEUTSCHE WELLE, 0400-0500 UTC

7

Frequency	Mode	Call Sign	Service / Times
7.150 MHz	AM		RADIO MOSCOW, (East Coast), 0000-0500 UTC, (November thru February)
7.155	AM		RADIO FREE EUROPE, (Latvian), 1500-2200 UTC
7.160	AM		VOICE OF AMERICA, (Service to Middle East), 0430-0530 UTC
	AM		BRITISH BROADCASTING CORPORATION, 1700-2030 UTC
7.165	AM		RADIO FREE EUROPE (Romanian), 0400-0700, 1800-2200 UTC
7.170	AM		VOICE OF AMERICA, (English to Middle East/Europe service) 0400-0700 UTC, (Russian to USSR) 1600-1800, 1900-2300 UTC
7.171	LSB		SLOW SCAN TELEVISION
7.175	AM		VOICE OF AMERICA, (Service to Europe), 1900-2100 UTC
	AM		RADIO MOSCOW, (West Coast), 0730-0900 UTC, (November thru February)
7.180	AM		RADIO FREE EUROPE, (Azerbaijan) 0000-0100 UTC, (Armenian) 0100-0200 UTC, (Georgian), 0200-0300 UTC, (Armenian), 0300-0400 UTC, (Belorussian), 0400-0500 UTC
	AM		BRITISH BROADCASTING CORP., 1300-1615 UTC

Frequency	Mode	Call Sign	Service / Times
7.190 MHz	AM		VOICE OF AMERICA, (Russian to USSR), 0300-0400 UTC
	AM		RADIO FREE EUROPE, (Polish), 0400-0600 UTC, (Romanian) 0700-0900 UTC, (Romanian), 1000-1500 UTC, (Polish) 1500-2200 UTC, (Kirghiz), 2300-2400 UTC
	AM		RADIO AFRICA, EQUATORIAL GUINEA, 2000-2300 UTC
	VOICE		SPY STATION, SPANISH FEMALE VOICE, SENDS 4-DIGIT TRAFFIC, 0430 UTC
7.200	AM		VOICE OF AMERICA, (English to Middle East/Europe service) 0400-0600, (Service to China), 0000-0100, 2000-2400 UTC,
7.205	AM		VOICE OF AMERICA, (English to Middle East/Europe service), 0100-0300 UTC, (English to VOA Europe) 0100-0300 UTC
7.210	AM		VOICE OF AMERICA, (Service to Central Asia), 0000-0100 UTC
	AM		RED CROSS BROADCASTING SERVICE, (English), (Sundays), 1100-1130, (Mondays), 1700-1730 UTC
7.215	AM		ISLAMIC REPUBLIC OF IRAN BROADCASTING, (English Service), 1130-1230 UTC
7.220	AM		THE VOICE OF IRAN, (English Service), 1400-1500 UTC

Frequency	Mode	Call Sign	Service / Times
7.220 MHz	AM		RADIO FREE EUROPE, (Romanian), 0000-2400 UTC
7.225	AM		VOICE OF AMERICA, (Vietnamese to East Asia), 2230-2330 UTC (Khmer to East Asia), 2200-2230 UTC
	AM		RADIO DEUTSCHE WELLE, 0400-0500 UTC
7.230	AM		BRITISH BROADCASTING CORPORATION, 0400-0730, 1300 UTC
7.240	AM		VOICE OF AMERICA, (Service to USSR), 0200-0300 UTC
	AM		RADIO AUSTRALIA, 1100-2100 UTC
	AM		RADIO MOSCOW, (East Coast), 0300-0500 UTC, (November-February)
7.245	LSB		AMATEUR FAX
	AM		RADIO FREE EUROPE, (Romanian), 0000-0200, 0300-0600, 1800-2400 UTC
	AM		VOICE OF AMERICA, (Ukrainian to the USSR), 1600-1800 UTC (Service to Europe), 1430-1500 UTC
7.255	AM		RADIO FREE EUROPE, (Romanian), 0000-0200, 2100-2400 UTC
	AM		VOICE OF AMERICA, (Service to Middle East), 0430-0530 UTC
7.260	AM		VOICE OF AMERICA, (South East Asia), 2200-2300 UTC
	AM		RADIO MOSCOW, (West Coast), 0730-0900 UTC, (November thru February)

Frequency	Mode	Call Sign	Service / Times
7.270 MHz	AM		VOICE OF AMERICA, (Russian to USSR) 1800-2000, 2100-2400 UTC (Ukrainian to USSR), 0300-0500, 1600-1800 UTC
	AM		RADIO MOSCOW, (West Coast), 0530-0900 UTC, (November thru February)
7.275	AM		VOICE OF FREE CHINA, 2200-2300 UTC
	AM		RADIO SOUTH KOREA, (KBS), (English Service), 0600-0700 UTC, (Spanish Service), 1900-1945 UTC
	AM		RADIO ITALY, (English Service), 1935-1955 utc
7.280	AM		VOICE OF AMERICA, (Service to USSR), 2100-2200, (Caribbean) 1900-2000, (Middle East), 1700-1900, (South Asia), 1600-1700 UTC
7.285	AM		POLISH RADIO WARSAW, (English Service) 1400-1425, 1730-1755 (Spanish Service) 1430-1455, 2130-2155 UTC
7.290	LSB	W1AW	ARRL VOICE BULLETIN
7.295	AM		RADIO FREE EUROPE, (Turkmen), 0100-0200 UTC, (Romanian), 1500-1800 UTC, (Ukrainian) 1800-2300 UTC

Frequency	Mode	Call Sign	Service / Times
7.295 MHz	AM		VOICE OF AMERICA, (Serbo-Croatian to Europe), 0445-0500 UTC
7.303	FAX	JHM	TOKYO, JAPAN, WEATHER, 0000-2400 UTC
7.304	SSB	VJJ	ROYAL FLYING DOCTOR, AUSTRALIA, 0800-1700 UTC
7.310	AM		RADIO MOSCOW, (East Coast), 0130-0500 UTC, (November - February)
7.315	CW	KRH51	U.S. DEPT OF STATE, LONDON, GREAT BRITAIN CW AND 50, 75 BAUD RTTY
	AM	WHRI	NOBLESVILLE, IN, 0000-0800 UTC
7.324	USB		AIR MARS ANDREWS AFB, MARYLAND
7.325	AM		BRITISH BROADCASTING CORPORATION, 0000-0330, 1800-2200 UTC
	AM		VOICE OF AMERICA, (English to Middle East/Europe Service), 0600-0700 UTC
7.330	USB		SKYKING BROADCASTS
7.335	VOICE	CHU	OTTAWA CANADA, STANDARD TIME/FREQUENCY
7.340	SSB	VZK	ROYAL FLYING DOCTOR, AUSTRALIA, 0800-1600 UTC
7.345	AM		RADIO PRAGUE
	AM		RADIO MOSCOW, (West Coast), 0730-0900 UTC, (November thru February)
	AM		RADIO CZECHOSLOVAKIA, (English Service), 1800-1827, 1930-1957, 2100-2130, 2200-2225, 0000-0027, 0100-0130, 0300-0330, 0400-0430 UTC
7.355	AM	WRNO	NEW ORLEANS, 0000-0600 UTC

Frequency	Mode	Call Sign	Service / Times
7.355 MHz		AM	WYFR FAMILY RADIO NETWORK, 1100-1300 UTC, (English to Europe), 0600-0800, 2000-2200 UTC, (Spanish to Europe), 2200-2300 UTC
	AM	KNLS	THE NEW LIFE STATION, 1300-1400 UTC
	AM	WRNO	WRNO WORLDWIDE RADIO, 2300-0300 UTC
7.357	RTTY		U.S. ARMY MARS
7.35850	USB	AAE	HF/MARS RADIO STATION, FT SAM HOUSTON, TX
7.365	USB	NPG	NAVAL COMMUNICATIONS, STOCKTON, CA
	AM	KNLS	THE NEW LIFE STATION, 0800-0900 UTC
7.371	CW		SPY STATION, MORSE CODE, SENDS 5-DIGIT TRAFFIC, 2130 UTC
7.3725	USB	NAV	HQ NAVY-MARINE CORPS MARS RADIO STATION, CHELTENHAM, MD
7.375	USB		RADIO FOR PEACE INTERNATIONAL, 0000-1200 UTC
	AM		UNITED NATIONS RADIO, 2150-2200, 2345-0000 UTC, (Spanish Service), 1400-1415, 1530-1600 UTC
7.392	SSB	VJI	ROYAL FLYING DOCTOR, AUSTRALIA, 0800-1700 UTC
7.393	USB	NMN	COAST GUARD COMMUNICATIONS STATION, PORTSMOUTH, VA
7.395	AM	WRNO	0000-0055 UTC
	AM		CHRISTIAN SCIENCE MONITOR, 0300-0600 UTC, (Occasional Use Times) 0600-0800 UTC
	AM	WRNO	WRNO WORLDWIDE RADIO, 0300-0400 UTC
7.400	RTTY		IRAN NEWS AGENCY
7.401	ARQ		INTERPOL FREQUENCIES
7.403	FAX	ATP	WEATHER, NEW DELHI, INDIA
7.405	AM		VOICE OF AMERICA, 0100-1200 UTC
	AM		RADIO BEIJING, CHINA, (English Service), 1400-1600 UTC

Frequency	Mode	Call Sign	Service / Times
7.410 MHz	AM		THE VOICE OF ISRAEL, (English Service), 0500-0515 UTC
	SSB	VJJ	ROYAL FLYING DOCTOR, AUSTRALIA, 0800-1700 UTC
7.412	USB		NASA MISSION FREQ
7.415	AM/LSB		PIRATE STATIONS SEEM TO LIKE THIS FREQUENCY
	VOICE		SPY STATION, SPANISH FEMALE VOICE, SENDS 5-DIGIT TRAFFIC, 1030, 1100 UTC
7.416	AM		RADIO "THE VOICE OF VIETNAM", (English Service), 1100-1130
7.42185	USB		MILITARY SECURE VOICE
7.422	VOICE		SPY STATION, SPANISH FEMALE VOICE, SENDS 4-DIGIT TRAFFIC, 0000, 0200 0300, 0305, 2300 UTC 5-DIGIT TRAFFIC, 0400, 2245 UTC (WILL BROADCAST ON 5.930 AT SAME TIME)
7.435	AM		RADIO NEW YORK INTERNATIONAL, 0300-0600 UTC
	VOICE		SPY STATION, SPANISH FEMALE VOICE, SENDS 4-DIGIT TRAFFIC, 0600 UTC
7.438	RTTY		75 BAUD, BAUDOT RXREV OFF
7.442	USB	NKW	USN, DIEGO GARCIA
7.445	AM		VOICE OF AMERICA
7.453	FAX	AOK	WEATHER, ROTA, SPAIN
7.456	USB		U.S. NAVY MARS
7.457	USB		MILITARY MARS, (Air Force)
7.465	AM		THE VOICE OF ISRAEL, (English Service), 0000-0030, 0100-0130 0200-0230, 0500-0515 UTC
	SSB VJN		ROYAL FLYING DOCTOR, AUSTRALIA, 0800-1700 UTC
7.475	USB		WHISKEY 104
	AM		VOICE FOR A FREE IRAQ, 0400 UTC
	SSB VJI		ROYAL FLYING DOCTOR, AUSTRALIA, 0800-1700 UTC
7.478	AM		VOICE OF AMERICA (Relay link)

Frequency	Mode	Call Sign	Service / Times
7.480 MHz	AM		RED CROSS BROADCASTING SERVICE, (Mondays), 1310-1327 UTC
	AM		SWISS RADIO INTERNATIONAL, 1330-1400 UTC
7.490	AM	WWCR	NASHVILLE, TN, 0300-0800 UTC
7.500	RTTY		SPY STATION, BAUDOT, SENDS 5-DIGIT TRAFFIC, 0700 UTC
7.507	USB		NAVY HURRICANE WARNING
7.510	AM	KTBN	SALT LAKE CITY, UTAH, 0100 UTC
7.514	CW	RHA	OLA OLA OLA OLA DE RHA RHA RHA QSA ? QSA ? QTC 1 QSY ?? K
7.520	RTY	KRH50	U.S. DEPT OF STATE
	AM	WWCR	NASHVILLE, TN, 0100 UTC
7.527	VOICE		SPY STATION, SPANISH FEMALE VOICE, SENDS 5-DIGIT TRAFFIC, 0500 UTC
7.530	USB		U.S. COAST GUARD BOSTON, MA FAX NIK BOSTON, MASS.
7.532	AQR	OEQ	INTERPOL, VIENNA, AUT
	ARQ		INTERPOL FREQUENCIES
7.533	FAX	AXI	DARWIN, AUSTRALIA, WEATHER, 0800-2100 UTC
7.550	AM		RADIO SOUTH KOREA, (KBS), (English Service), 0800-0900, 2030-2130 UTC, (Spanish Service), 2215-2300 UTC
	SSB	VJO	ROYAL FLYING DOCTOR, AUSTRALIA, 0700-1700 UTC
7.56150	ASCII		98 BAUD, ASCII, RXREV OFF
7.565	SSB	VJC	ROYAL FLYING DOCTOR, AUSTRALIA 0700-1900 UTC
7.570	RTTY	KRH	U.S. EMBASSY, LONDON, ENGLAND
7.577	USB		U.S. COAST GUARD NEW ORLEANS, LA
7.58750	FAX	6VU	WEATHER, DAKAR
7.591	RTTY		100 BAUD, 6-BIT RTTY, RXREV ON
7.600	AM/CW		TIMING SIGNALS FROM ECUADOR (5 minutes before the hour)
7.603	RTTY		KUWAIT NEWS SERVICE
7.606	USB	VLB	ISRAELI MOSSAD INTELLIGENCE

Frequency	Mode	Call Sign	Service / Times
7.610 MHz	AM		RADIO EGYPT
7.620	AM		VOICE OF AMERICA
7.6275	CW	KWS78	QRA QRA QRA DE KWS78 KWS78 KWS78 QSX 7/10/14/18/23 K (Will transmit call once every minute)
7.635	USB		CIVIL AIR PATROL, SOUTHWESTERN, U.S.A. (Communicator Net)
7.64110	PAKT		MARS PACKET NET
7.648	AM		VOICE OF AMERICA
7.651	LSB/USB		VOICE OF AMERICA, 0300-0700 UTC, (Service Relay)
7.65350	AM		VOICE OF AMERICA
7.654	VOICE		SPY STATION, SPANISH FEMALE VOICE, SENDS 5-DIGIT TRAFFIC, 2100 UTC ENGLICH VOICE, 3-2 TRAFFIC, 2100 UTC
7.689	RTTY	TUH	RTTY 450 SHIFT AT 50 BAUD RYRYRYRYRY'S QJH1 QJH1 QJH1 DE TUH TUH TUH TFC TESTING (TUH location is Abidjani, Ivory Coast)
7.691	CW	3FTV	DE 3FTV (CW Net)
7.69250	CW	CMU967	RMIZ DE CMU967 QRU ? QSV QRA 1 QTC ? K, (Cuban Military), 0450, 0800 UTC (Out-Station heard on 11113.50 sending to CMU967)
7.70540	CW	NMN	CQ CQ NMN/NAM/NAK/NAR/GXS/AOK WEATHER QRU K
7.710	FAX	VFF	FROBISHER BAY, WEATHER
7.720	RTTY		U.S. ARMY MARS
7.722	AM		VOICE OF AMERICA
7.724	CW	KRH50	QRA QRA QRA DE KRH50 KHR50 KRH50 QSX 5/7/11/13/16/20 ? QSX 5/7/11/13/16/20 K, (U.S. Embassy, London, Great Britain) (Will transmit call every minute)
7.725	VOICE		SPY STATION, SPANISH FEMALE VOICE, SENDS 5-DIGIT TRAFFIC, 0900, 1000 1030 UTC (WILL BROADCAST ON 10.324 AT SAME TIME)
7.734	VOICE		SPY STATION, SPANISH FEMALE VOICE, SENDS 5-DIGIT TRAFFIC, 0600 UTC
7.742	USB		NASA MISSION FREQ

Frequency	Mode	Call Sign	Service / Times
7.755 MHz	USB	MKG	R.A.F. LONDON
7.765	USB		NASA MISSION FREQ
7.767	AM		VOICE OF AMERICA
7.77650	RTTY	RCW	RUSSIAN
7.790	VOICE		SPY STATION, ENGLISH FEMALE VOICE, SENDS 5-DIGIT TRAFFIC, 0300, 0600, 0900 UTC
7.800	RTTY		IRAN NEWS AGENCY
7.803	SSB VJI		ROYAL FLYING DOCTOR, AUSTRALIA, 0800-1700 UTC
7.804	ARQ	OEM	INERPOL, VIENNA, AUT
7.831	USB		LOOKING GLASS
7.832	USB		AIR USAF WASHINGTON, DC
	ARQ		INTERPOL FREQUENCIES
7.845	VOICE		SPY STATION, SPANISH FEMALE VOICE, SENDS 5-DIGIT TRAFFIC, 0400, 0610 UTC
	CW		SPY STATION, SPANISH FEMALE VOICE, SENDS 5-DIGIT TRAFFIC, 0830 UTC
7.858	VOICE		SPY STATION, SPANISH FEMALE VOICE, SENDS 5-DIGIT TRAFFIC, 0330, 0530 UTC
7.860	CW		SPY STATION, MORSE CODE, SENDS 5-DIGIT TRAFFIC, 0230 UTC
7.862	VOICE		SPY STATION, SPANISH FEMALE VOICE, SENDS 5-DIGIT TRAFFIC, 0500, 0530, 0930, 1000, 1030 UTC
7.870	CW		SPY STATION, MORSE CODE, SENDS 5-DIGIT TRAFFIC, 0830 UTC
7.880	FAX	DDK	WEATHER, QUICKBORN, GERMANY
7.887	VOICE		SPY STATION, SPANISH FEMALE VOICE, SENDS 5-DIGIT TRAFFIC, 0300, 0400, 0500, 0530, 0600, 0700, 0800, 0820, 1000, 1900, 2100, 2200 UTC ENGLISH FEMALE VOICE, SENDS 5-DIGIT TRAFFIC, 0300, 0400, 0500, 0900 UTC
7.888	VOICE		SPY STATION, ENGLISH FEMALE VOICE, SENDS 5-DIGIT TRAFFIC, 0300, 0500 UTC
7.890	VOICE		SPY STATION, SPANISH FEMALE VOICE, SENDS 5-DIGIT TRAFFIC, 0400, 0500, 0530 UTC

Frequency	Mode	Call Sign	Service / Times
7.906 MHz	ARQ		INTERPOL FREQUENCIES
7.969	ARQ		INTERPOL FREQUENCIES
7.975	SSB	VJY	ROYAL FLYING DOCTOR, AUSTRALIA, 0800-2400 UTC
7.910	FAX	RWJ	RUSSIAN WEATHER
7.918	USB		CIVIL AIR PATROL
7.924	PKT		CIVIL AIR PATROL PACKET NET
7.960	RTTY		IRAN NEWS AGENCY
7.983	USB		SKYKING BROADCASTS
7.993	FAX	NPM	PEARL HARBOR, HI, WEATHER
7.99810	RTTY	FDY	80 BAUD, RXREV OFF

TIME CONVERSION CHART

U.T.C.	PST	PDST MST	MDST CST	CDST EST	EDST
0:00	4 pm	5 pm	6 pm	7 pm	8 pm
1:00	5 pm	6 pm	7 pm	8 pm	9 pm
2:00	6 pm	7 pm	8 pm	9 pm	10 pm
3:00	7 pm	8 pm	9 pm	10 pm	11 pm
4:00	8 pm	9 pm	10 pm	11 pm	Midnight
5:00	9 pm	10 pm	11 pm	Midnight	1 am
6:00	10 pm	11 pm	Midnight	1 am	2 am
7:00	11 pm	Midnight	1 am	2 am	3 am
8:00	Midnight	1 am	2 am	3 am	4 am
9:00	1 am	2 am	3 am	4 am	5 am
10:00	2 am	3 am	4 am	5 am	6 am
11:00	3 am	4 am	5 am	6 am	7 am
12:00	4 am	5 am	6 am	7 am	8 am
13:00	5 am	6 am	7 am	8 am	9 am
14:00	6 am	7 am	8 am	9 am	10 am
15:00	7 am	8 am	9 am	10 am	11 am
16:00	8 am	9 am	10 am	11 am	Noon
17:00	9 am	10 am	11 am	Noon	1 pm
18:00	10 am	11 am	Noon	1 pm	2 pm
19:00	11 am	Noon	1 pm	2 pm	3 pm
20:00	Noon	1 pm	2 pm	3 pm	4 pm
21:00	1 pm	2 pm	3 pm	4 pm	5 pm
22:00	2 pm	3 pm	4 pm	5 pm	6 pm
23:00	3 pm	4 pm	5 pm	6 pm	7 pm

Frequency	Mode	Call Sign	Service / Times
8.000 MHz	VOICE	JJY	TIMING SIGNAL FROM JAPAN
8.006	ARQ		INTERPOL FREQUENCIES
8.014	SSB	VZK	ROYAL FLYING DOCTOR, AUSTRALIA, 0800-1600 UTC
8.030	USB		CUBA MILITARY
8.0345	USB		CUBA MILITARY
8.035	SSB	VZZ	ROYAL FLYING DOCTOR, AUSTRALIA 0800-1600 UTC
8.037	CW		SPY STATION, MORSE CODE, SENDS 5-DIGIT TRAFFIC, 0300, 0500, 0600 UTC
8.038	ARQ		INTERPOL FREQUENCIES
8.040	FAX	GFA	BRACKNELL, UNITED KINGDOM 0000-2400 UTC
8.042	CW	BZI	QRA DE BZI, (People Republic of China,Weather)
8.045	ARQ	OEQ	INTERPOL, VIENNA, AUT
	ARQ		INTERPOL FREQUENCIES
8.050	RTTY		IRAN NEWS AGENCY
8.0515	FEC		WEATHER REPORTS
8.0535	ARQ	WOO	CQ CQ CQ DE WOO WOO WOO DIGITAL SELECTIVE CALLING (DSC)
8.056	VOICE		SPY STATION, SPANISH FEMALE VOICE, SENDS 5-DIGIT TRAFFIC, 0300, 0400, 0600 UTC
8.070	VOICE		SPY STATION, SPANISH FEMALE VOICE, SENDS 4-DIGIT TRAFFIC, 0100, 0300 UTC (WILL BROADCAST ON 10.324 AT SAME TIME)
8.0775	FAX	SMA	WEATHER, SWEDEN
8.080	FAX	NAM	NORFOLK, VA, WEATHER
8.0847	USB	KMI	AMTOR RXREV ON CQ CQ CQ DE KMI KMI KMI DIGITAL CALLING (DSC)
8.087	FEC		WEATHER REPORTS
	ARQ	KMI	WEATHER REPORTS
	VOICE		SPY STATION, SPANISH FEMALE VOICE, SENDS 5-DIGIT TRAFFIC, 0300, 0600 0700 UTC (WILL BROADCAST ON 7.887 AT SAME TIME)
8.088	USB	C3C	MILITARY NET
8.089	ARQ		INTERPOL FREQUENCIES
8.0892	CW	NMN	CQ CQ CQ DE NMN/NAM/NRK/NAR/GXH/ AOK QRU

8

Frequency	Mode	Call Sign	Service / Times
8.0894 MHz	ARQ	KMI	CQ CQ CQ DE KMI KMI KMI GM/GE, WX AND STATION INFO ARE SENT AT 20 PAST ODD UTC HOURS ATT&T KMI/WOM/WOO CURRENT LIST OF STATIONS THAT HAVE RADIO TELEPHONE CALLS WAITING FOR THEM ARE AS FOLLOWS:
8.094	VOICE		SPY STATION, SPANISH FEMALE VOICE, SENDS 5-DIGIT TRAFFIC, 0600, 0605 UTC
8.097	ARQ		INTERPOL FREQUENCIES
8.101	USB		SKYKING BROADCASTS
8.110	AM		VOICE OF AMERICA
8.111	CW		SPY STATION, MORSE CODE, SENDS 5-DIGIT TRAFFIC, 0200, 0230, 0330 UTC (WILL REPEAT TRAFFIC FROM 6.785)
8.113	USB		INTERNATIONAL SSB RADIO TELEPHONY, (Receive on 8713.00)
8.11575	RTTY		75 BAUD 6-BIT RTTY RXREV ON
8.1175	CW	BMB	VVV VVV VVV CQ CQ CQ DE BMB BMB BMB FREQ 3641/5909 /8118/13560 KHZ HR WX
8.118	FAX	BAF	BEIJING, PEOPLES REPUBLIC OF CHINA, WEATHER
8.119	CW	E9T	SENDS CODED TRAFFIC PLUS RTTY CODED TRAFFIC, ALSO TRANSMITS ON 14761 AT SAME TIME
8.1219	FAX	BAF	WEATHER, PEKING, CHINA
8.122	ARQ		INTERPOL
	VOICE		SPY STATION, SPANISH FEMALE VOICE, SENDS 5-DIGIT TRAFFIC, 0600 UTC
8.127	USB	CIO2	ISRAELI MOSSAD
8.135	VOICE		SPY STATION, SPANISH FEMALE VOICE, SENDS 5-DIGIT TRAFFIC, 0400, 0430, 0500, 0600, 0730, 0800, 1130 UTC
	CW		SPY STATION, MORSE CODE, SENDS 5-DIGIT TRAFFIC, 0230, 1130 UTC
8.144	SSB	VJO	ROYAL FLYING DOCTOR, AUSTRALIA, 0700-1700 UTC
8.1466	FAX	IMB	WEATHER, ROME, ITALY
8.149	CW	OVG	VVV VVV VVV OVG 5/8/12 ZKR 6 7 10 11 MC/S
8.150	CW	NRV	U.S. COAST GUARD

Frequency	Mode	Call Sign	Service / Times
8.159 MHz	CW	P7X	DE P7X, (Sends 5-digit coded traffic)
8.165	SSB	VNZ	ROYAL FLYING DOCTOR, AUSTRALIA, 0800-1700 UTC
8.171	SSB	VNZ	ROYAL FLYING DOCTOR, AUSTRALIA 0800-1700 UTC
8.180	USB	C1U23	U.S. MILITARY (0800 UTC)
8.185	FAX	FP188	PARIS, FRANCE, WEATHER
	VOICE		SPY STATION, SPANISH FEMALE VOICE, SENDS 5-DIGIT TRAFFIC, 0500, 0600, 0630, 0800, 0900 UTC (WILL ALSO BROADCAST ON 13.374)
	CW		SPY STATION, MORSE CODE, SENDS 5-DIGIT TRAFFIC, 0300, 0330, 0340, 0430 UTC (ALSO USES 11.632 AND 12.214)
8.190	USB		NONE VOICE 5 DIGIT NUMBERS
8.193	CW		SPY STATION, MORSE CODE, SENDS 5-DIGIT TRAFFIC, 0330 UTC
8.2048	USB		MARITIME UNITS
8.216	USB		SPANISH VOICE
8.225	USB		SPANISH VOICE
8.226	USB		MARITIME UNITS
8.2352	USB		RUSSIAN AEROFLOT
8.2507	USB		RUSSIAN AEROFLOT
8.257	SSB		MARITIME SAFETY AND DISTRESS
8.264	USB		INTERNATIONAL SSB RADIO TELEPHONY, (Receive on 8788.00)
8.270	VOICE		SPY STATION, ENGLISH FEMALE VOICE, SENDS 5-DIGIT TRAFFIC, 0300 UTC
8.277	USB		RUSSIAN AEROFLOT
8.282	USB		INTERNATIONAL SSB RADIO TELEPHONY, (Receive on 8806.00)
8.28430	USB		RUSSIAN AEROFLOT
8.29110	USB		SHIP TO SHIP, SHIP TO SHORE COMMUNICATIONS.
8.2942	USB		SHIP TO SHIP, SHIP TO SHORE COMMUNICATIONS.
8.29430	USB		RUSSIAN AEROFLOT
8.311	ARQ	6OF	
8.32622	ALIST		100 BAUD ALIST RXREV ON
8.34673	ALIST		100 BAUD ALIST RXREV ON
8.350	AM		RADIO BAGHDAD, 2100-2300 UTC
8.3627	CW	JRC/WCC	CALLING CQ

Frequency	Mode	Call Sign	Service / Times
8.364 MHz	CW		INTERNATIONAL LIFEBOAT, LIFERAFT AND SURVIVAL CRAFT
8.3636	USB		JAPANESE MARITIME AGENCY
	CW	OBC3	DE OBC3 QRU QTC QSV K
8.3644	USB		JAPANESE MARITIME AGENCY
	CW	56JT	VDS VDS VDS DE 56JT 56JT 56JT QTC QSA 3 MSG K
8.3655	CW	C0G7	POSSIBLE CUBAN MILITARY
8.3687	CW	ZL6W	5LW'S DE ZL6W QSA NIL EE
8.3716	USB		JAPANESE MARITIME AGENCY
8.3738	CW	C6IW2	VIS DE C6IW2, (Ship at sea passing traffic)
8.38130	CW		WX TRAFFIC
8.4175	CW	KFS	DE KFS KFS KFS SITOR SELCAL 1.01094 QSX 4 6 8 12 AND 16 MHZ K
	CW	WNU	DE WNU SELCAL 1109 8417/8377
8.4185	ARQ	WLO	(Receive on 08378.50)
8.419	ARQ	WLO	(Receive on 08379.00)
8.420	ARQ	VCS	
8.4205	ARQ	CBV	
8.421	ARQ	FFT	
	ARQ	CBV	
8.421	ARQ	WLO	(Receive on 08381.00)
8.4215	ARQ	WLO	(Receive on 08381.50)
8.4225	ARQ	KPH	
	ARQ	YUR	
8.423	ARQ	KPH	
8.4235	ARQ	VIS	
	ARQ	WLO	(Receive on 08383.50)
8.4245	ARQ	WCC	
8.4255	ARQ	KLB	
8.426	ARQ	NMC	
8.427	ARQ	WCC	
8.428	ARQ	NMN	
8.429	ARQ	PCH	
	ARQ	WLO	(Receive on 08389.00)
8.4295	ARQ	RMO	
8.430	ARQ	NMO	
8.4305	ARQ	DHS	DE DHS
8.433	ARQ	WLO	
8.435	ARQ	OST	
8.437	CW	4XZ	VVV DE 4XZ 4XZ BT BT

Frequency	Mode	Call Sign	Service / Times
8.439 MHz	CW		SPY STATION, MORSE CODE, SENDS 5-DIGIT TRAFFIC, 0500 UTC
8.4408	CW	VCS	VVV VVV VVV CQ DE VCS VCS VCS QSX 4 6 AND 8 MHZ CHNL 3/4/7/8
8.445	CW	KFS	CQ DE KFS KFS KFS/B QSX 8 12 16 22 MHZ K
8.4455	CW	WLO	DE WLO 3 OBS ? AMVERS ? QSX 4 6 8 12 22 25R071 MHZ ANS CH/6 K
8.448 MHz	CW	LPO	WEATHER/AMVER
8.4525	CW	VAI	CQ CQ CQ DE VAI VAI VAI QSX 4/6/8/12/16 MHZ CH4/5 OBS/AMVER/QRJ/WESTREG ? VAI SITOR SELCALL 1.00581 QRU ? K
8.457	USB	NMA	USCG, MIAMI, FLA
8.459	FAX	NOJ	KODIAK, AK, WEATHER
8.460	CW	CBA	CQ CQ DE CBA CBA QSX 8 MHZ COMMON AND CH 1 K
	CW	PPJ	WEATHER/AMVER
8.461	CW	ZSC	WEATHER/AMVER
8.4624	CW	CKN	NAWS DE CKN II ZKR F1 2386 4156.6 6254 8314 12465.5 16642 22178 KHZ AR
8.4637	CW	JOU	CQ CQ CQ DE JOU JOU JOU QSX 8 MHZ K
8.467	FAX	JJC	JAPAN WEATHER
8.46930	CW	XFL	CQ CQ CQ DE XFL XFL XFL
8.4718	CW	NMN	CQ CQ CQ DE NMN NMN NMN QRU ? K CQ CQ CQ DE NMN NMN NMN QRU ? QSX 8/12/16 MHZ ITU CHANS 4/5/6 BT AS OF 01 JULY 91 HERE IS NEW ITU CHANS/FREQ NMN GUARDS BT COMMON CH-4 8369/12553.5/16738 KHZ. CH-5 8367/12551/16735 KHZ.CH-6 8367.5/12551.5/16735.5 KHZ. HERE IS THE A1A WORKING BANDS (FREQ CHANS IN .5 KHZ INCREMENTS) 8 MHZ 8342 THRU 8376 KHZ.12 MHZ 12422 THRU 12476.5 KHZ16 MHZ 16619 THRU 16683 KHZ. BT DE NMN QRU ? K
8.47275	CW	WLO	DE WLO 3 OBS ? AMVERS ? QSX 6 8 12 16 25R071 MHZ NW ANS CH/6 K
8.474	CW	WLO	DE WLO 2 OBS ? AMVERS ? QSX 4 6 8 12 16 22 25R172 MHZ NW ANS C 3/4 K

Frequency	Mode	Call Sign	Service / Times
8.47785MHz	CW	FUF	VVV DE FUF
8.478	CW	VHP	VVV VVV VVV DE VHP VHP VHP 2/3/4/5/6/8 AR
8.4797	CW	JCU	CQ CQ CQ DE JCU JCU JCU QSX 8 MHZ K
8.484	CW	DAN	CQ CQ CQ DE DAN DAN DAN 8 CG 16 CG K QSX GROUP CHANNEL 8370.5 AND 16738.5 KHZ K
8.485	USB	SYN	ISRAELI MOSSAD INTELLIGENCE
8.4862	CW	4XO	CQ DE 4XO QSX 12 C K
8.487	CW	XSG	WEATHER/AMVER
8.490	USB	NMF	USCG, BOSTON, MA
8.492	FAX	NPM	PEARL HARBOR, HI, WEATHER
8.4925	CW	PPR	VVV DE PPR PPR PPR QSX 8 MHZ K
8.49485	FAX	GYA	NORTHWOOD, UNITED KINGDOM, 0000-2400 UTC
8.4965	CW	CLA	CQ CQ DE CLA CLA QSX C/6 8368/12552/16736 TX 8573/12673 16961 QSW CLA 20/32/41/50 K
8.4985	CW	SAG	CQ DE SAG2/4/6 QSX GROUP CH 4 AND COMMON CH 4 OR 4812 MHZ BT FOR QRJ QSX 420 801 1203 AR
8.5025	CW	PPL	VVV VVV VVV DE PPL PPL PPL QSX 4/8 AND 22 MHZ K
	CW	NIK	INTERNATIONAL ICE PATROL
8.504	CW	ZLB	DE ZLB2/4/5/6/7 ZLB2/4/5/6/7 QSX 4/8/12/16/22 MHZ CHLS 3/4/10 BT
8.5045	CW	ZLW	DE ZLW ZLW QSX 4 8 12 16 22 MHZ CHL 3 4 1 0
8.505	FAX	AOK	WEATHER, MADIRD, ROTA, SPAIN
8.510	CW	FFL	CQ CQ CQ DE FFL3/FFL4 FFL3/FFL4 QSX 6/8 MHZ BT
8.5125	CW	DAL	VVV DE DAL DAL
8.514	FEC		WEATHER REPORTS
8.515	CW	5AT	VVV VVV VVV CQ CQ CQ DE 5AT 5AT 5AT
	CW	VJZ	WEATHER/AMVER
8.51535	CW	GKC	DE GKC
8.5205	CW	PPO	VVV DE PPO PPO PPO QSX CHANNELS 4/5 ON 22 MHZ QSW 22450 KHZ QSX CHANNELS 2/3
8.521	CW	VIS26	VVV VVV VIS26 K VIS26 K
8.522	CW	CBV	DE CBV QSX CH 1/3/4 ON 4 6 8 12 MHZ AUTOTLX 4 8 12 16 BT K

Frequency	Mode	Call Sign	Service / Times
8.5234 MHz	CW	JOR	CQ CQ CQ DE JOR JOR JOR QSX 8 MHZ K
8.5235	CW	FFL	V CQ CQ CQ DE FFL2/3/4 FFL2/3/4 QSX 4 AND 8 MHZ CHANNEL 1 TO 6 K
8.5255	CW	WNU33	CQ CQ CQ DE WNU33 WNU33 WNU33 QSX 4 6 8 12 16 MHZ OBS ?
8.52675	CW	LGW	CQ CQ DE LGW LGU LGB LGN LGJ LGX LFX QSX 4 G 6 CG 8 CG 12 CG 16 CG AND 16740.7
8.5279	FAX	CTU	WEATHER, MONSANTO
8.5325	ARQ	WLO	(Weather and Traffic List)
8.5385	CW	6WW	VVV DE 6WW
8.5465	CW	GKA	VVV VVV DE GKA
8.5476	CW	JFA	CQ CQ CQ DE JFA JFA JFA K
8.550	CW	PWZ	WEATHER/AMVER
8.552	CW	CTP	DE CTP CTP QSX 4 6 8 12 MHZ AR
8.553	RTTY	UXN	USSR RTTY
8.5575	CW	SPE	DE SPE QRL AS
8.558	CW	KFS	CQ DE KFS KFS KFS/A QSX 8 12 16 AND 22 MHZ K
8.5598	CW	GKB	DE GKB
8.5625	CW	PCH	DE PCH40 8 AS
8.5658	CW	D3E41	CQ DE D3E41/51/52 QSX 6 AND 8 MHZ C.6 ON RTF CH 4 2182 11621 AND 2207 AR
8.5695	CW	XFM	VVV DE XFM XFM XFM QRU ? 4/8/12 CQ CQ CQ DE XFM XFM XFM QRU ? QRU ? 8/12
8.570	CW	WNU43	CQ CQ CQ DE WNU43 WNU43 WNU43 QSX 4 6 8 12 16 MHZ
8.571	CW	JNA	CQ CQ CQ DE JNA JNA JNA
8.5745	CW	LGW	CQ CQ CQ DE LGW LGB LFN LGJ LGX QSX 4 CG 8 CG 12 CG AND 16 CG
8.575	CW	NMC	CQ CQ CQ DE NMC NOR NOC QRU ?
8.573	CW	CLA	CQ CQ DE CLA CLA QSX C/6 8364R4/12546R6/16728R8 TX 8573R0/12873R5/16761 QSW CLA 20/31/32/41 K
8.583	CW	KLB	CQ CQ CQ DE KLB KLB KLB QSX 4 6 8 12 16 AND 22 MHZ OBS ? SITOR SELCAL 1113 AR K
8.584	CW	DHS	CQ DE DHS DHS QSX 8/12 MHZ CH 8 QSJ 8584 KHZ AR
8.58535	CW	WCC	VVV VVV DE WCC WCC BT OBS ? QSX 6 8 12 MHZ K

Frequency	Mode	Call Sign	Service / Times
8.589 MHz	CW	HPP	VVV VVV CQ CQ CQ DE HPP HPP HPP QTC ? AMVER ? OBS ? BT QSW RTG 500 KHZ / 8.589 / 12.699 / 16.869 MHZ QSK CH 5 / 6 / 9 BT RTTY CH 08 / 12 MHZ CH 33 / 16 MHZ BT
8.595	CW	UIK	WEATHER/AMVER
8.5964	CW	CKN	VVV VVV VVV DE CKN CKN CKN C13E C13E C13E
8.5975	CW	VIP03	VVV DE VIP03 QSX 5 6 16
8.601	CW	CWA	CQ CQ CQ DE CWA CWA QSX 4/6/8/12/16 MHZ CH/6/16/18 AND 22 MHZ C 3/4/9 K
8.6015	CW	ZLO	DE ZLO ZAY A1A 6 8 12 ZNI 1A 6 12 ZNI 16 J2B 8 12 J7B 8 16 AR
8.6045	CW	ZRH	DE ZRH QSX 4 X 8 X 12 16 X X DE ZRH QSX
8.606	CW	ZRQ	VVV VVV VVV ZRQ 4/6
8.610	CW	WMH	WEATHER/AMVER
8.6165	CW	WPP	WEATHER/AMVER
8.617	FAX	JJC	TOKYO, JAPAN, WEATHER
8.6182	CW	KPH	VVV VVV VVV DE KPH KPH QSX 4 6 8 12 16 22 K
8.6205	CW	PPO	WEATHER/AMVER
8.622	CW	PCH41	DE PCH41 8 K DE PCH41 8 AS DE PCH41 8 K
8.6245	CW	FUM	VVV DE FUM
8.625	CW	GYU	DE GYU
8.6285	CW	NOJ	CQ CQ CQ DE NOJ NOJ TFC LIST NIL BT QTC ? OBS/AMVER/ GOVT/FISHERIES ? QSX 8368.8/8364/500 KHZ DE NOJ AR
8.6308	CW	WCC	VVV VVV DE WCC WCC BT OBS ? QSX 6 8 12 MHZ K
8.6342	CW	PPR	VVV DE PPR PPR PPR QSX 8 MHZ K
8.636	CW	HLW	CQ CQ CQ DE HLW HLW HLW QSX 8 MHZ K
8.63930	CW	DAM	VVV DE DAM DAM
8.640	CW	E	"E" MARKER
8.642	CW	KPH	W2 VVV DE KPH QSX 4 6 8 12 16 22 CQ DE KFH QSX 425 KHZ OR HF QRU ? OBS ? AMVER ?
8.646	FAX	WWD	LAJOLLA, CA, WEATHER
8.6467	CW	FUJ	VVV DE FUJ
8.648	CW	G23B	VVV VVV VVV DE G23B G23B
8.649	CW	ICB	VVV DE ICB ICB ICB K 4 8 12 MHZ

Frequency	Mode	Call Sign	Service / Times
8.650 MHz	CW	NMO	CQ CQ CQ DE NMO NMO QRU AMVER ? QSX 8/12 MHZ AMVER CHNL 5/6/11 ITU 22 MHZ AMVER CHNL 3/4 ITU. QLH 8650/ 12889.5 KHZ AND 22476 KHZ DE NMO NMO NMO QRU ? K
8.6525	CW	OST	VVV DE OST4/42 OST5/52 ANS 8 OR 22 MHZ C
8.654	CW	JCS	CQ CQ CQ DE JCS JCS JCS QSX 8 MHZ K
8.656	CW	IAR	VVV DE IAR IAR IAR K 4 8 12
8.659	CW	WLO	DE WLO 3 OBS ? AMVERS ? QSX 4 6 8 12 16 22 25R172 MHZ NW ANS C 3/4 DE WLO 3 OBS ? AMVERS ? QSX 4 6 8 12 16 22 25R172 MHZ ANS C 5/6 K
8.665	CW	KLC	CQ CQ DE KLC KLC KLC TFC LIST QSW 4 8 4 AND IIF 4/6/8/12/16/22 MHZ BT (Galveston, Texas)
8.6665	CW	FUG	VVV VVV VVV DE FUG 4/6/8 FUG 4/6/8 FUG 4/6/8 FAAG FAAG FAAG
8.6705	CW	IAR	VVV DE IAR IAR IAR K 4 8 12
8.678	ARQ	URD	
8.6795	CW	IQX	VVV VVV VVV DE IQX IQX IQX
8.680	CW	WSC	CQ CQ DE WSC WSC QSX 6 8 12 MHZ OBS ? DE WSC K
8.6815	CW	EAD	DE EAD2/EAD3 QSX 6/8 MHZ CG AR K
8.682	FAX	NMC	SAN FRANCISCO, CA, WEATHER
8.6839	ARQ	LGB	LGB TLX
8.688	CW	9VG	VVV DE 9VG36 18688 KHZ (Singapore)
8.689	CW	WNU53	CQ CQ CQ DE WNU53 WNU53 WNU53 QSX 4 6 8 12 MHZ OBS ?
8.6895	CW	ZSC	CQ CQ DE ZSC ZSC ZSC AR K
8.6932	CW	SVD4	DE SVD4
8.694	CW	Q	"Q" CHANNEL MARKER
	CW	PJC	CQ CQ DE PJC PJC TFC COMING ON 4334 AND 8694 KHZ AS
8.6945	CW	4XO	CQ DE 4XO QSX 12 C K
	CW	D3E41	CQ DE DE D3E41/51/52/62 QSX 6 8 AND 12 MHZ TFC ON RTF CH 421 871 1621 BT AND 2207 AR
8.69635	CW	CFH	VVV VVV VVV DE CFH CFH CFH C13L C13L C13L
8.698	CW	7TF6	CQ CQ CQ DE 7TF6 7TF6 7TF6 QSX 8367/8368/8369 KHZ K

Frequency	Mode	Call Sign	Service / Times
8.700 MHz	CW	YUR3	VVV DE YUR3/5/7 CH 3/4/7 K
8.7005	CW	HKB	CQ CQ CQ DE HKB HKB HKB QSX ON 8.364/12.546/16.728 MHZ K
8.70375	CW	KFS	DE KFS KFS KFS SITOR SELCAL 1.01094 QSX 4 6 8 2 AND 16 MHZ K
8.7048	CW	WNU	CW DE WNU SEL 1109 8705.5/8344.5
8.7055	ARQ	WNU	
8.706	CW	KFS	DE KFS KFS KFS SITOR SELCAL 1.01094 QSX 4 6 8 12 AND 16 MHZ
8.7065	ARQ	NMC	
	CW	JOS	CQ CQ CQ DE JOS JOS JOS QSX 8 MHZ K
8.7085	ARQ	WPP	
8.7105	ARQ	NMF	
8.711	ARQ	KPH	
8.712	ARQ	VIS65	
		WCC	
8.71223	ARQ	NMC	
	ARQ	KPH	
	ARQ	KLB	
8.713	ARQ	HPP	
8.714	ARQ	NMC	
8.71424	ARQ	NMN	
	ARQ	WCC	
8.718	ARQ	NMO	
		NMN	
8.7185	ARQ	WLO	
		KFS	
8.7227	USB		MARITIME UNITS
8.7259	USB		MARITIME UNITS
8.72830	CW	KMI	DE KMI
8.737	USB		VANCOUVER COAST GUARD
8.740	USB	DTUU	AT&T HIGH SEAS
8.7435	USB	WOO	AT&T HIGH SEAS
8.744	CW	KPH	VVV DE KPH KPH QSX 4 6 8 12 16 MHZ
8.764	USB		NATIONAL WEATHER SERVICE, (Sometimes Computer Voice)
8.7654	USB	NMC	U.S. COAST GUARD
8.7778	USB		U.S. MILITARY (Will change call sign daily)
8.790	USB	WLO	MARITIME UNITS
8.805	CW		SPY STATION, MORSE CODE, SENDS 5-DIGIT TRAFFIC, 0000 UTC
8.825	USB		INTERNATIONAL AIRLINES

Frequency	Mode	Call Sign	Service / Times
8.828 MHz	USB		AERONAUTICAL WEATHER REPORTING
8.842	CW		RUSSIAN AIRLINE AEROFLOT
8.843	USB	CA125	U.S. NAVY
8.846	USB		INTERNATIONAL AIRLINES
8.847	USB		KUWAIT CITY AIR
8.874	VOICE		SPY STATION, SPANISH FEMALE VOICE, SENDS 5-DIGIT TRAFFIC, 0600, 0605 UTC
8.879	USB		INTERNATIONAL AIRLINES
8.885	VOICE		SPY STATION, SPANISH FEMALE VOICE, SENDS 5-DIGIT TRAFFIC, 0400 UTC
8.888	USB		CUBA AIR LINES
8.891	USB		INTERNATIONAL AIRLINES
8.912	USB		U.S. CUSTOMS
8.9135	USB		U.S. CUSTOMS, (Secure Voice)
8.921	USB		INTERNATIONAL AIRLINES
8.925	VOICE		SPY STATION, ENGLISH FEMALE VOICE, SENDS 5-DIGIT TRAFFIC, 0730 UTC
8.927	USB		CUBA AIR LINES
8.939	USB		RUSSIAN AIR LINES
8.959	USB		SAUDI AIR LINES
8.963	USB		NIGERIA AIR LINES
8.964	USB		USAF TACTICAL AIR COMMAND
	CW	6WW	VVV DE 6WW
8.967	USB		US AIR FORCE
8.9685	USB		US AIR FORCE
8.984	USB		U.S. COAST GUARD AIRCRAFT
8.9768	USB		INTERNATIONAL AIRLINES
8.989	USB		MILITARY AIRLIFT COMMMAND
	USB		USAF GLOBAL CONTROL AND COMMAND
8.993	USB		MILITARY AIRLIFT COMMAND

DATE	FREQ.	MODE	TIME	STATION	SIGNAL	COMMENTS	SWL SENT	SWL REC'D

Frequency	Mode	Call Sign	Service / Times
9.0015 MHz	USB	NWC	U.S. NAVY
9.00775	RTTY		75 BAUD 6-BIT
9.0055	ARQ		INTERPOL
9.006	USB		SKYKING BROADCASTS
9.014	USB		USAF AWACS EARLY WARNING
	USB		USAF TACTICAL AIR COMMAND
9.017	USB		LOOKING GLASS
9.022	AM		ISLAMIC REPUBLIC OF IRAN BROADCASTING, (English Service), (Beaming Europe), 1930-2030 UTC, (Spanish Service), 0130- 0230, 0530-0630, 2030-2130 UTC
9.023	USB		SKYKING BROADCASTS
	USB		USAF NORAD HQ.
9.027	USB		SKYKING BROADCASTS
	USB		LOOKING GLASS
9.03595	USB		USAF TACITAL
9.043	USB		NASA MISSION FREQ
9.0449	FAX	5YE	WEATHER, NAIROBI, AFRICA
9.050	AM		RADIO CANADA INTERNATIONAL, 0515-0600 UTC, (English Service)
	CW	NMO	VVV VVV VVV DE NMO NPM NMO NPM NMO NPM FCM-2 NUKO NUKO NUKO (COMMSTA, Honolulu, HI)
9.057	USB		SKYKING BROADCASTS
	USB		LOOKING GLASS
9.08467	RTTY		76 BAUD BAUDOT RXREV ON
9.08857	RTTY		76 BAUD BAUDOT RXREV ON
	FAX	NPM	WEATHER REPORTS, HONOLULU, HAWAII
9.090	VOICE		SPY STATION, ENGLISH FEMALE VOICE, SENDS 3-2 DIGIT TRAFFIC, 2100, 2300 UTC
9.0966	RTTY		75 BAUD BAUDOT RXREV ON
9.10158	RTTTY		75 BAUD
9.102	VOICE		SPY STATION, ENGLISH FEMALE VOICE, SENDS 3-2 DIGIT TRAFFIC, 0400, 0600 UTC 4-DIGIT TRAFFIC, 0500, 0600 UTC 5-DIGIT TRAFFIC, 0600, 0800 UTC

Frequency	Mode	Call Sign	Service / Times
9.1045 MHz	ARQ		INTERPOL
9.105	ARQ		INTERPOL FREQUENCIES
9.1136	RTTY		45 BAUD
9.120	CW	WGY912	FEDERAL EMERGENCY MANAGEMENT ADMINISTRATION
9.130	USB	EZI	ISRAELI MOSSAD INTELLIGENCE
	USB		USAF OVERSEAS LINK
	VOICE		SPY STATION, SPANISH FEMALE VOICE, SENDS 5-DIGIT TRAFFIC, 0700, 0720, 0800 UTC
9.132	USB		NASA MISSION FREQ
	ARQ		INTERPOL FREQUENCIES
9.135	FAX	JBK4	TOKYO, JAPAN, 0000-2400 UTC
9.154	VOICE		SPY STATION, SPANISH FEMALE VOICE, SENDS 5-DIGIT TRAFFIC, 0300,0500, 0600, 0700, 0730, 0800, 0810, 0830, 0900, 1000 UTC
	CW		SPY STATION, MORSE CODE, SENDS 5-DIGIT TRAFFIC, 0200, 0300, 1000 UTC (WILL ALSO USE 9.322)
9.157	FAX	WLO	MOBILE, AL, WEATHER
9.185	ARQ		INTERPOL FREQUENCIES
9.200	LSB		100 BAUD ARQ, INTERPOL
	ARQ		INTERPOL FREQUENCIES
9.203	FAX	GFE	BRACKNELL, UNITED KINGDOM
9.205	USB		US FORCES NATO
9.207	USB		US NAVY MARS
9.209	USB	NKW	USN, DIEGO GARCIA
9.2175	CW	BT02	VVV BT01 DE BT02 K, (Possible China military) (Will send 4-digit cut numbers)
9.220	USB		SKYKING BROADCASTS
9.222	VOICE		SPY STATION, SPANISH FEMALE VOICE, SENDS 4-DIGIT TRAFFIC, 0200, 0205 0230, 0300, 0330 UTC (WILL ALSO USE 12.156 AND 16.540)
9.230	FAX	RTB	RUSSIAN WEATHER
	VOICE		SPY STATION, SPANISH FEMALE VOICE, SENDS 4-DIGIT TRAFFIC, 0400, 0405, 0600 UTC
9.234	USB		SKYKING BROADCASTS

Frequency	Mode	Call Sign	Service / Times
9.235 MHz	VOICE		SPY STATION, SPANISH FEMALE VOICE, SENDS 5-DIGIT TRAFFIC, 0600 UTC 4-DIGIT TRAFFIC, 0400 UTC
9.238	VOICE		SPY STATION, SPANISH FEMALE VOICE, SENDS 5-DIGIT TRAFFIC, 0500 UTC
9.240	VOICE		SPY STATION, SPANISH FEMALE VOICE, SENDS 5-DIGIT TRAFFIC, 0600 UTC
9.246	VOICE		SPY STATION, SPANISH FEMALE VOICE, SENDS 5-DIGIT TRAFFIC, 0300, 0400 UTC
9.255	VOICE		SPY STATION, SPANISH FEMALE VOICE, SENDS 5-DIGIT TRAFFIC, 0400 UTC
9.265	VOICE		SPY STATION, ENGLISH FEMALE VOICE, SENDS 5-DIGIT TRAFFIC, 0700, 1000 UTC
9.285	ARQ		INTERPOL FREQUENCIES
9.322	CW		SPY STATION, MORSE CODE, SENDS 5-DIGIT TRAFFIC, 0300, 0330, 1000, 1500 UTC
9.331	VOICE		SPY STATION, SPANISH FEMALE VOICE, SENDS 5-DIGIT TRAFFIC, 0500, 1000 UTC
9.342	RTTY		KUWAIT NEWS SERVICE
9.350	USB		VOICE OF AMERICA
9.353	FAX	RNR	RUSSIAN WEATHER
9.3585	FAX	OXT	DENMARK, 0000-1800 UTC
9.365	VOICE		SPY STATION, SPANISH FEMALE VOICE, SENDS 5-DIGIT TRAFFIC, 0200, 0230, 0300, 0305, 0400 UTC
9.383	FAX	NPN	APRA, GUAM, WEATHER
9.395	FAX	NPM	PEARL HARBOR, HI, WEATHER
	FAX	NKW	WEATHER
9.400	AM		VOICE OF IRAQ, 0230 UTC
9.402	VOICE		SPY STATION, ENGLISH FEMALE VOICE, SENDS 5-DIGIT TRAFFIC, 0530, 0700 UTC
9.410	AM		BRITISH BROADCASTING CORPORATION, 0300-0730, 1500-2330 UTC
9.425	AM		CHRISTIAN SCIENCE MONITOR, 1200-1400 UTC
9.430	VOICE		SPY STATION, ENGLISH FEMALE VOICE, SENDS 5-DIGIT TRAFFIC, 0600 UTC
9.435	AM		THE VOICE OF ISRAEL, (English Service), 0000-0030, 0100-0130, 0200-0230, 0500-0515 UTC

Frequency	Mode	Call Sign	Service / Times
9.435 MHz	AM	KOL	ISRAEL RUSSIAN SECTION, 0500-0540, 2300-2355 UTC
	VOICE		SPY STATION, ENGLISH FEMALE VOICE, SENDS 5-DIGIT TRAFFIC, 0530, 1000 UTC
9.438	FAX	JMJ3	TOKYO, JAPAN, WEATHER
9.44025	RTTY		75 BAUD 6-BIT RTTY RXREV ON
9.445	AM		THE VOICE OF TURKEY, (English Service), 0355-0500, 2300-2400 UTC
9.455	AM		VOICE OF AMERICA, (English to Caribbean service), 0000-0100 UTC (Brazil), 2300-2400 UTC (Latin America), 0100-0200 UTC
	AM		CHRISTIAN SCIENCE WORLD SERVICE, 0200-1200 UTC (Saturaday & Sunday), 2005-2155 UTC, (Spanish Service), (Sabado, Domingo), 0305-0355, 0505-0555, 0705-0755 UTC
9.457	FAX	ZKLF	WELLINGTON, NEW ZEALAND, WEATHER, 0000-2400 UTC
9.463	RTTY		JORDAN NEWS AGENCY
9.465	AM	WHRI	INTERNATIONAL RADIO, 0000-0900, 1300-1500 UTC
	AM		WORLD SERVICE BOSTON, MA, 2200-2400 UTC
	AM		CHRISTIAN SCIENCE MONITOR, 2200-2255 UTC
	AM		VOICE OF AMERICA, (Service to Latin America), 0100-0400 UTC
9.470	AM		RADIO MOSCOW WORLD SERVICE, (English Service) 0200-0500 UTC
9.475	AM		NORTHERN MARIANAS, 2000 UTC
	AM		RADIO CAIRO, EGYPT, 0200-0330 UTC
	AM		WHRI, NOBLESVILLE, TN, 2300-2400 UTC

Frequency	Mode	Call Sign	Service / Times
9.475 MHz	AM		CHRISTIAN SCIENCE MONITOR, 1200-1255 UTC
9.495	AM		CHRISTIAN SCIENCE MONITOR, 1000-1255 UTC
9.503	USB		SKYKING BROADCASTS
9.505	AM		RADIO FREE EUROPE, (Latvian), 1500-2200 UTC
	AM	WYFR	FAMILY RADIO NETWORK, 0100-0500 UTC
	AM		VOICE OF AMERICA, (Service to South East Asia), 1130-1230, (Europe), 0300-0330 UTC
	AM		RADIO JAPAN 1400-1600, 1700-1800, 1900-1930 UTC
	AM		RADIO MOSCOW, (West Coast), 0500-0900 UTC, (November thru February)
	AM		UNITED NATIONS RADIO, (Monday, Saturday), 2100-2130, 1845-1900, 2015-2030 UTC
9.510	AM		VOICE OF AMERICA, (Vietnamese to East Asia), 2230-2330 UTC
9.515	AM		BRITISH BROADCASTING CORPORATION, WORLD SERVICE, 1100-1400 UTC, (Spanish Service), 0300-0430 UTC
	AM		RADIO SOUTH KOREA, (KBS), (Spanish Service), 1900-1945 UTC
9.520	AM		RADIO FREE EUROPE, (Romanian) 0000-2400 UTC

Frequency	Mode	Call Sign	Service / Times
9.525 MHz	AM		VOICE OF AMERICA, (English to Pacific service) 1900-2000 UTC (Spanish to Cuba), 2300-0600 UTC (Turkish to Europe/Middle East), 2000-2100 UTC
	AM		POLISH RADIO WARSAW, (English Service) 1300-1355, 1730-1755 UTC (Spanish Service) 1430-1455, 2130-2155 UTC
9.530	AM		VOICE OF AMERICA, (English to Africa serivce), 0600-0700 UTC (English to Middle East/Europe service) 2230-2400 UTC (Urdu to South Asia) 0130-0200 UTC
	AM		RADIO MOSCOW WORLD SERVICE, (English Service) 0200-0300 UTC
	AM		RADIO MOSCOW, (English Service, East Coast) 2330-0400 UTC, (Sept 1 - Sept 28)
	AM		CHRISTIAN SCIENCE MONITOR, (Northeast Asia), 1405-1555 UTC (Saturaday & Sunday), 1405-1555 UTC
	AM		RADIO MOSCOW, (East Coast), 0030-0500 UTC
	AM		RADIO EXTERIOR DE ESPANA, (Spanish Service) 0200-0500 UTC
9.535	AM		RADIO JAPAN 1400-1500 UTC
	AM		SWISS RADIO INTERNATIONAL, 0730-0800 UTC

Frequency	Mode	Call Sign	Service / Times
9.535 MHz	AM		RADIO DEUTSCHE WELLE, (Radio Germany), 0300-0400, 0500-0600 UTC
9.540	AM		VOICE OF AMERICA, (Service to South Asia), 0100-0300 UTC (Middle East), 2000-2100 UTC
	AM		RADIO CZECHOSLOVAKIA, (English Service), 0000-0027, 0100-0130, 0300-0330, 0400-0430 UTC, (Spanish Service), 0200-0227 UTC
	AM		POLISH RADIO WARSAW, (English Service) 1500-1525, 1600-1625 UTC
9.545	AM		RADIO DEUTSCHE WELLE, (Radio Germany), 0300-0500 UTC
	AM		VOICE OF AMERICA, (Service to China), 0000-0100, 2000-2400
	AM		RADIO BERLIN GERMANY, 2100 UTC
9.550	AM	WYFR	FAMILY RADIO NETWORK, (Spanish Service), 1000-1100 UTC
9.555	AM		RADIO FREE EUROPE, (Bulgarian), 1300-1500, 1800-2100 UTC
	AM		RADIO FREE AFGHANISTAN, 0230-0330 UTC
	AM	WYFR	FAMILY RADIO NETWORK, (Spanish Service), 0800-1000 UTC
9.560	AM		THE VOICE OF JORDAN, (English Service), 1400-1740 UTC
	AM		VOICE OF ETHIOPIA, 1500-1550 UTC

Frequency	Mode	Call Sign	Service / Times
9.560 MHz	AM		RED CROSS BROADCASTING SERVICE, (English), 0740-0757 UTC (Also broadcasts on, 13685, 17670, 21695 MHZ)
	AM		SWISS RADIO INTERNATIONAL, 0200-0230, 0830-0900 UTC
	AM		RADIO FINLAND, (English Service), 0245-0345 UTC
9.565	AM		RADIO FREE EUROPE, (Uzbek), 1600-1700 UTC, (Ukrainian), 1700-2300 UTC
	AM		RADIO DEUTSCHE WELLE, 0400-0500 UTC
	AM		VOICE OF AMERICA, (Service to Latin America), 0000-0030, 2330-2400 UTC, (USSR), 0200-0500 UTC, (Africa), 0500-0700, (North Africa), 0730-0800 UTC, (USSR), 2000-2100 UTC
	AM		ORGANIZATION OF AMERICAN STATES, All programing in English except for Spanish broadcasts 2345-0030 UTC
9.570	AM		VOICE OF AMERICA, (English to Africa service) 2000-2100 UTC
	AM		RADIO SOUTH KOREA (KBS), (English Service), 1215-1315, 1400-1500 UTC, (Spanish Service), 1015-1100 UTC
9.575	AM		ISLAMIC REPUBLIC OF IRAN BROADCASTING, (English Service) (East Asia), 1130-1230 UTC

Frequency	Mode	Call Sign	Service / Times
9.575 MHz	AM		VOICE OF AMERICA, ALSO ON 06035.00,15115.00 (English to Africa service) 0300-2000 UTC (Russian to USSR) 1900-2200 UTC
	AM		RADIO ITALY, (English Service), 0100-0120 UTC
9.580	AM		VOICE OF AMERICA, (Service to North Africa), 0730-0800 UTC
	AM		BRITISH BROADCASTING CORP., 0000-0230 UTC
	AM		RADIO AUSTRALIA, 0830-2100 UTC
9.585	AM		VOICE OF AMERICA, (Russian to USSR) 1700-2100 UTC, (Europe), 1430-1500 UTC, (Caribbean), 1900-1930 UTC
	AM	HCJB	THE VOICE OF THE ANDES (English Service) 0700-0830
9.590	AM		VOICE OF AMERICA, (Spanish to Cuba) 1200-1400 UTC, (South Asia), 1600-1700 UTC, (Caribbean), 1000-1200 UTC
	AM		BRITISH BROADCASTING CORPORATION, 0000-0300, 2100-2400 UTC
	AM		RADIO NETHERLANDS INTERNATIONAL, (English Service), 0330-0430 UTC, (Spanish Service), 0430-0525 UTC
	AM		RADIO NORWAY INTERNATIONAL, 1300 AND 2100 UTC

Frequency	Mode	Call Sign	Service / Times
9.595 MHz	AM		RADIO FREE EUROPE, (Latvian), 0230-0300 UTC
	AM		RADIO JAPAN, 0500-0900 UTC
9.600	AM		BRITISH BROADCASTING CORPORATION, 0500-0530, 0600-0630, 0700-0730, 0800-0830 , 1800-2200 UTC
	AM		RADIO MOSCOW, (English Service, East Coast) 0000-0400 UTC, (Sept 1 - Sept 28)
	AM		RADIO MOSCOW WORLD SERVICE, (English Service) 1100-1200 UTC
9.605	AM		VATICAN RADIO, VATICAN CITY, 0030-0130 UTC
	AM		VOICE OF AMERICA, (Service to Europe), 0400-0500 UTC, (Caribbean), 0200-0230 UTC, (Europe), 0300-0400 UTC
	AM		RADIO NORWAY INTERNATIONAL, 0100-0300 UTC
	AM	WYFR	FAMILY RADIO NETWORK, (Spanish Service), 1100-1200 UTC
	AM		RADIO CZECHOSLOVAKIA, (English Service), 1800-1827, 2100-2130, 2200-2225 UTC
9.610	AM		VOICE OF AMERICA, (Service to South Asia), 1230-1330 UTC, (Indo-China), 1100-1230, 2200-2330 UTC
	AM		BRITISH BROADCASTING CORP., 0400-0430 UTC
9.615	AM		VOICE OF AMERICA, (Uzbek to the USSR), 0000-0100 UTC

Frequency	Mode	Call Sign	Service / Times
9.615 MHz	AM		RADIO NORWAY INTERNATIONAL, 0100-0300 UTC
9.620	AM		VOICE OF AMERICA, (Vietnamese to East Asia) 1230-1330 UTC
	AM		RADIO EXTERIOR DE ESPANA, (Spanish Service) 0900-1900 UTC
9.625	AM		RADIO FREE EUROPE, (Tajik), 0100-0200 UTC, (Ukrainian), 0200-0600 UTC, (Russian), 0600-1500 UTC, (Ukrainian), 1500-2300 UTC
9.630	USB		NATIONAL OCEANIC AND ATMOSPHERIC ADMINISTRATION (NOAA) THRUSDAY AT 0750 UTC
9.635	AM		RADIO FREE EUROPE, (Tatar-Bashkir), 0400-0600 UTC
	AM		VOICE OF AMERICA, (Urdu to South Asia), 0100-0130 UTC
	AM		RADIO CANADA INTERNATIONAL 1200-1300 UTC
9.640	AM		BRITISH BROADCASTING CORP. 0500-0730 UTC
	AM		RADIO JAPAN 1900-1930 UTC (Oceania)
	AM		RADIO SOUTH KOREA, (KBS), (Spanish Service), 0130-0230 UTC
	AM		RADIO DEUTSCHE WELLE, (Radio Germany), 0300-0400 UTC
9.645	AM		VOICE OF AMERICA, (English to South Asia), 1400-1800 UTC

Frequency	Mode	Call Sign	Service / Times
9.645 MHz	AM		RADIO NORWAY INTERNATIONAL, 2400-0100, 0300-0500 UTC
	AM		RADIO JAPAN 1900-1930 UTC, (Africa)
9.650	AM		RADIO MOSCOW, (English Service)
	AM		VOICE OF AMERICA, (Service to Europe), 1600-2030 UTC
	AM		RADIO SOUTH KOREA, (KBS), (English Service), 1130-1200 UTC
	AM		SWISS RADIO INTERNATIONAL, (English Service), 0000-0030, 0200-0230 UTC
9.655	AM		RADIO NORWAY INTERNATIONAL, 1700 UTC
	AM		RADIO MOSCOW WORLD SERVICE, (English Service) 1300-1600 UTC
9.660	AM		RADIO FREE EUROPE, (Kazak), 0000-0200 UTC, (Ukrainian), 0200-0600 UTC
	AM	KNLS	THE NEW LIFE STATION, 1300-1400 UTC
9.665	AM		THE VOICE OF TURKEY, (English Service), 2300-2400 UTC
	AM		RADIO BEIJING, CHINA, (English Service), 1200-1300 UTC
9.670	AM		VOICE OF AMERICA, (English to Middle East/Europe service), 0500-0600 UTC (Slovene to Europe), 0430-0500 UTC (Ukrainian to the USSR), 0300-0500 UTC (Turkish to Europe/Middle East), 2000-2100 UTC

Frequency	Mode	Call Sign	Service / Times
9.670 MHz	AM		RADIO DEUTSCHE WELLE, (Radio Germany), 0500-0600 UTC
9.675	AM		RADIO INDIA, (English Service), 0630 UTC
	AM		RADIO CAIRO, EGYPT, 0200-0300 UTC
9.680	AM		RADIO FREE EUROPE, (Azerbaijan), 0000-0100 UTC, (Armenian) 0100-0200 UTC, (Georgian) 0200-0300 UTC, (Armenian), 0300-0400 UTC, (Azerbaijan) 0400-0500 UTC, (Russian), 1700-2200 UTC, (Kazak) 2300-2300 UTC
	AM		VOICE OF FREE CHINA, (English Service), 0200-0400 UTC
	AM	WYFR	FAMILY RADIO NETWORK, (English to Europe), 0600-0800
9.685	AM		THE VOICE OF TURKEY, (English Service), 2300-2400 UTC
	AM		RADIO MOSCOW, (English Service, East Coast) 2300-0300 UTC (Aug 1 - Sept 28), (Sept - March), 0300-0500 UTC
	AM		RADIO MOSCOW WORLD SERVICE, (English Service) 0200-0500, 2100-2200 UTC

Frequency	Mode	Call Sign	Service / Times
9.690 MHz	AM		VOICE OF AMERICA, (Russian to USSR) 1600-1800, 1900-2100 UTC (Uzbek to the USSR), 0000-0100 UTC
	AM		RADIO BEIJING, CHINA, (English Service), 0300-0400 UTC
	AM		RADIO DEUTSCHE WELLE, (Radio Germany), 0500-0600 UTC
	AM		BRITISH BROADCASTING CORPORATION, WORLD SERVICE, (Spanish Service), 1100-1130, 1300-1330 UTC
9.695	AM		VOICE OF CAMBODIA, PHNOM-PENH, 0000 UTC
	AM		RADIO FREE EUROPE, (Hungarian), 1500-2200 UTC
	AM		ISLAMIC REPUBLIC OF IRAN BROADCASTING, (English Service), (Middle East), 1130-1230 UTC
	SSB	VJI	ROYAL FLYING DOCTOR, AUSTRALIA, 0800-1700 UTC
9.700	AM		BRITISH BROADCASTING CORP.
	AM		VOICE OF AMERICA, (English to Middle East/Europe service), 0500-0600, 1600-2200 UTC (English to South Asia), 1500-1800 UTC
	AM		RADIO NEW ZEALAND INTERNATIONAL, 0630-1210 UTC (Daily)
9.705	AM		THE KINGDOM OF SAUDI ARABIA, (English Service)
	AM		THE VOICE OF IRAN, (English Service), 1130-1225 UTC
	AM		RADIO MOSCOW WORLD SERVICE, 1700 UTC

Frequency	Mode	Call Sign	Service / Times
9.705 MHz	AM		RADIO FREE EUROPE, (Polish), 0400-0600, 1500-2200 UTC (Russian) 0700-1400 UTC
	AM		RADIO ETHIOPIA, 0400-0500 UTC
	AM		VOICE OF AMERICA, (Urdu to South Asia), 0100-0130 UTC
	AM		RADIO DEUTSCHE WELLE, (Radio Germany), 0300-0400 UTC
	AM	WYFR	FAMILY RADIO NETWORK, 1300-1500 UTC, (Spanish Service) 0500-0600 UTC
9.710	AM		RADIO AUSTRALIA, 1100-1530 UTC
	AM		RADIO LITHUANIA, 0600-0800 UTC
	AM		RADIO ITALY, (English Service) 1935-1955 UTC
9.715	AM		VOICE OF AMERICA, (English to Middle East/Europe service) 0400-0600 UTC, (Central Asia), 0000-0100 UTC
	AM		RADIO FREE EUROPE, (Turkmen) 0100-0200 UTC
	AM	WYFR	FAMILY RADIO NETWORK, (Spanish Service), 0000-0500 UTC
9.720	AM		THE KINGDOM OF SAUDI ARABIA, (English Service)
	AM		RADIO MOSCOW, (English Service, East Coast) 2300-0400 UTC (Sept 1 thru Sept 28)

Frequency	Mode	Call Sign	Service / Times
9.725 MHz	AM		RADIO FREE EUROPE, (Romanian), 0400-2200 UTC
9.732	AM		RADIO "THE VOICE OF VIETNAM", (English Service), 1100-1130 UTC
9.740	AM		BRITISH BROADCASTING CORPORATION, 0500-0730, 1030-1400 UTC
	AM		VOICE OF AMERICA, (Service to Central Asia), 0000-0100 UTC, (Middle East), 0300-0530 UTC
9.745	AM	HCJB	QUITO, ECUADOR, 0100-1100 UTC
	AM		VOICE OF AMERICA, (Service to East Asia), 2130-2200 UTC
	AM		ISLAMIC REPUBLIC OF IRAN BROADCASTING, (Spanish Service), 0530-0630 UTC
	AM	HCJB	THE VOICE OF THE ANDES (English Service) 0030-0430, 0730-1130 UTC
9.750	AM		RADIO SOUTH KOREA, (KBS), (English Service), 1215-1315 UTC
	AM		RADIO FREE EUROPE, (Russian) 0000-0100 UTC, (Uzbek) 0100-0300 UTC
	AM		BRITISH BROADCASTING CORPORATION, 0900-1630 UTC
	AM		RADIO MOSCOW, (East Coast), 0400-0500 UTC, (Nov thru Feb)

Frequency	Mode	Call Sign	Service / Times
9.755 MHz	AM		RADIO CANADA INTERNATIONAL, 2200-2230, 2330-0030 (Saturday & Sunday) 2330-0100 UTC
	AM		VOICE OF AMERICA, (Service to East Asia), 2130-2200 UTC
	AM		RADIO MOSCOW WORLD SERVICE, (English Service) 1300-1700 UTC
9.760	AM		VOICE OF AMERICA, (English to Pacific service) 1100-1500 UTC
	AM		(English to Middle East/Europe service), 1630-2200 UTC (Spanish to American Republics), 0100-0400 UTC
	AM		BRITISH BROADCASTING CORP., 0700-1530 UTC
9.765	AM		VOICE OF FREE CHINA, TAIWAN, 0200-0400 UTC
	AM		RADIO DEUTSCHE WELLE, 0400-0500 UTC
	AM		ISLAMIC REPUBLIC OF IRAN BROADCASTING, (Spanish Service), 0130-0230 UTC
	AM		RADIO MOSCOW, (East Coast), 0300-0500 UTC, (Nov thru Feb)
	AM	HCJB	THE VOICE OF THE ANDES (Spanish Service) 1030-1500
9.770	AM		RADIO BEIJING, CHINA, (English Service), 0000-0100, 0300-0400 UTC
	AM		VOICE OF AMERICA, (English to Pacific service) 2200-0100 UTC
	AM		RADIO AUSTRALIA, 1430-1530 UTC
	AM		RADIO DEUTSCHE WELLE, (Radio Germany), 0300-0400 UTC

Frequency	Mode	Call Sign	Service / Times
9.775 MHz	AM		VOICE OF AMERICA, (English to American Republics service), 0000-0230 UTC, (Africa), 0430-0500 UTC
	AM		RADIO NETHERLANDS INTERNATIONAL, (Spanish Service), 1200-1225 UTC
9.780	AM		RADIO YEMEN, 0800-1400 UTC
9.785	AM	KVOH	VOICE OF HOPE, (English Service) 0400-0800 UTC
9.793	USB		USAF NORAD HQ.
9.795	AM		THE VOICE OF TURKEY (English Service)
	AM		RADIO MOSCOW, (West Coast), 0630-0900 UTC, (Nov thru Feb)
	AM		SWISS RADIO INTERNATIONAL, (English Service), 2200-2230 UTC
9.815	AM		VOICE OF AMERICA, (English to American Republics service), 0000-0230 UTC
9.820	AM		RADIO SOUTH KOREA, (KBS), (English Service), 1600-1700 UTC
9.825	AM		RADIO MOSCOW, (West Coast), 0730-0900 UTC, (Nov thru Feb)
	AM		BRITISH BROADCASTING CORPORATION, WORLD SERVICE, (Spanish Service), 0000-0200, 0300-0430 UTC
9.835	AM		RADIO BUDAPEST, HUNGARY, 0100 UTC

Frequency	Mode	Call Sign	Service / Times
9.840 MHz	AM		CHRISTIAN SCIENCE RADIO, 0400-0800 UTC
	AM		VOICE OF AMERICA, (Service to Latin America), 0100-0400 UTC
	AM		RADIO "THE VOICE OF VIETNAM", (English Service), 1000-1030, 1230-1300, 1330-1400, 1600-1630, 1800-1830, 1900-1930, 2030-2100, 2330-2400 UTC, (Spanish Service), 1100-1130, 2000-2030 UTC
9.850	AM		CHRISTIAN SCIENCE RADIO, 0000-0200 UTC
	AM	WYFR	FAMILY RADIO NETWORK, (English to Europe), 0500-0600 UTC
9.8525	AM		VOICE OF FREE CHINA, (English Service) (Europe Beaming can be heard in North America), 2200-2300 UTC
9.860	AM		RADIO AUSTRALIA, 1430-2100 UTC
9.867	RTY	YIZ74	IRAQ NEWS AGENCY
9.870	AM		CHRISTIAN SCIENCE MONITOR, (Occasional Use Frequency) 0400-0600 UTC, (Spanish Service), (Sabado, Domingo), 0605-0655 UTC
	AM		RADIO AUSTRIA INTERNATIONAL, (English Service), 0130-0200, 0330-0400, 2130-2200 UTC, (Spanish Service), 0030-0100, 0230-0300, 2230-2300 UTC
	AM		KOREAN BROADCASTS SERVICE, 1715-1730, 2045-2100 UTC
	AM		RADIO MOSCOW, (East Coast), 0000-0100 UTC, (Nov thru Feb)

Frequency	Mode	Call Sign	Service / Times
9.875 MHz	AM		RADIO MOSCOW WORLD SERVICE, 1300-1700 UTC
	AM		RADIO AUSTRIA INTERNATIONAL, 0330-0400 UTC, (Spanish Service), 0230-0330 UTC
9.885	AM		RED CROSS BROADCASTING SERVICE, (English), 0310-0327 UTC 1710-1727 UTC (Thursdays)
	AM		SWISS RADIO INTERNATIONAL, (English Service), 2200-2230, 0000-0030, 0200-0230, 0400-0430 UTC, (Spanish Service), 2230-2300, 0030-0100, 0230-0300 UTC
9.895	AM		RADIO NETHERLANDS INTERNATIONAL, (Spanish Service), 2330-0025, 0230-0325 UTC
	AM		RADIO MOSCOW, (West Coast), 0500-0700 UTC, (Nov thru Feb), (East Coast), 0430-0500 UTC (Nov thru Feb)
9.900	AM		RADIO CAIRO, EGYPT, 2100-2300 UTC
9.915	AM		BRITISH BROADCASTING COMPANY, 0000-0600, 2200-2400 UTC
9.921	ARQ		INTERPOL FREQUENCIES
9.930	ARQ		INTERPOL FREQUENCIES
9.958	VOICE		SPY STATION, SPANISH FEMALE VOICE, SENDS 4-DIGIT TRAFFIC, 0200, 0230, 0500 UTC (ALSO WILL USE 6.840)
9.968	FAX	JMH	TOKYO, JAPAN, WEATHER, 0000-2400 UTC
9.973	CW	KRH51	U.S. DEPT OF STATE, LONDON, GREAT BRITAIN

Frequency	Mode	Call Sign	Service / Times
10.000 MHz	VOICE	WWV	INTERNATIONAL STANDARDS TIME FREQUENCY
	VOICE	JJY	TIMING SIGNAL FROM JAPAN
10.015	USB		NATIONAL HURRICANE SERVICE
10.04630	CW	4XZ	VVV DE 4XZ 4XZ 4XZ BT BT (Has been copied sending 5 digit coded traffic)
10.051	USB		AERONAUTICAL WEATHER REPORTING
10.100	**SSB/CW**	**START**	**OF AMATEUR RADIO 30 METER BAND (Ends 10150.00)**
10.113	FAX	BAF	BEIJING, PEOPLES REPUBLIC OF CHINA, WEATHER
10.115	FAX	5YE	WEATHER, NAIROBI, AFRICA
10.123	FAX	SUU	WEATHER, CAIRO, EGYPT
10.12835	RTTY		50 BAUD
10.1315	RTTY		50 BAUD 6 BIT RTTY
10.150	AM		RADIO EGYPT
10.1508	RTY	SUA	RY'S, 50 BAUD, BAUDOT, RXREV OFF
10.157	RTTY		KUWAIT NEWS SERVICE
10.1625	RTY	YIL70	IRAQ NEWS AGENY
10.166	USB		U.S. NAVY
10.200	USB	MKG	R.A.F., LONDON
10.225	AM	PPN	WEATHER, BRASILIA, BRAZIL
10.235	AM		VOICE OF AMERICA
	VOICE		SPY STATION, SPANISH FEMALE VOICE, SENDS 4-DIGIT TRAFFIC, 0200, 0205, 0230 UTC
10.2375	AM		VOICE OF AMERICA
10.246	USB		SKYKING BROADCAST, ALSO USES SECURE VOICE
10.247	VOICE		SPY STATION, ENGLISH FEMALE VOICE, SENDS 4-DIGIT TRAFFIC, 0300, 0305, 0700, 0705 UTC
10.24825	RTTY		75 BAUD 6-BIT RTTY RXREV ON
10.258	USB		SKYKING BROADCASTS
10.260	CW		SPY STATION, MORSE CODE, SENDS 5-DIGIT TRAFFIC, 1000 UTC
10.270	VOICE		SPY STATION, SPANISH FEMALE VOICE, SENDS 5-DIGIT TRAFFIC, 0200, 0230, 0400 UTC
10.283	RTTY		85 BAUD

Frequency	Mode	Call Sign	Service / Times
10.295 MHz	ARQ		INTERPOL, 100 BAUD
	ARQ		INTERPOL FREQUENCIES
10.305	USB		NASA MISSION FREQ
10.310	USB		NASA MISSION FREQ
10.324	VOICE		SPY STATION, SPANISH FEMALE VOICE, SENDS 4-DIGIT TRAFFIC, 0100, 1030 UTC (WILL ALSO USE 7.725 AND 8.070
10.36515	RTTY		75 BAUD
10.382	AM		VOICE OF AMERICA
10.385	AM		VOICE OF AMERICA
10.390	LSB		100 BAUD ARQ, INTERPOL
	ARQ		INTERPOL FREQUENCIES
10.391	ARQ		INTERPOL FREQUENCIES
10.4155	CW		VVV VVV VVV DE JJD/JJD2 JJD/JJD2 JJD/JJD2 CQ CQ CQ DE JJD/JJD2 JJD/JJD2
10.42639	ALIST		75 BAUD TDM 2:8 RXREV OFF
10.445	VOICE		SPY STATION, SPANISH FEMALE VOICE, SENDS 5-DIGIT TRAFFIC, 0400, 0600 UTC
10.452	USB		SKYKING BROADCASTS
10.467	VOICE		SPY STATION, SPANISH FEMALE VOICE, SENDS 4-DIGIT TRAFFIC, 0400, 0700 UTC
	CW		SPY STATION, MORSE CODE, SENDS 5-DIGIT TRAFFIC, 0705 UTC
10.470	CW	KKN50	QRA QRA QRA DE KKN50 KKN50 KKN50 QSX 6/10/12/15 K (Will transmit call once every minute)
10.485	VOICE		SPY STATION, SPANISH FEMALE VOICE, SENDS 5-DIGIT TRAFFIC, 0400 UTC
10.494	RTTY	6WW	FRENCH MILITARY 75 BAUD
10.510	USB		SKYKING BROADCASTS
	VOICE		SPY STATION, ENGLISH FEMALE VOICE, SENDS 5-DIGIT TRAFFIC, 0400, 0500 UTC
10.515	RTTY		SWISS RADIO INTERNATIONAL, (Swiss Radio broadcasts via RTTY five times a day in English, German and French), 0030-0130, 0200-0300 UTC, (beamed to North and South America)
10.530	VOICE		SPY STATION, ENGLISH FEMALE VOICE, SENDS 3-2 DIGIT TRAFFIC, 0400, 1300, 2200 UTC
10.53435	RTTY		75 BAUD BAUDOT RXREV OFF (weather traffic)
10.536	FAX	CFH	HALIFAX, NOVA SCOTIA

Frequency	Mode	Call Sign	Service / Times
10.540 MHz	VOICE		SPY STATION, SPANISH FEMALE VOICE, SENDS 5-DIGIT TRAFFIC, 0400, 0500 UTC
	CW		SPY STATION, MORSE CODE, SENDS 5-DIGIT TRAFFIC, 0330 UTC
10.545	AM		VOICE OF AMERICA
10.553	FAX	AXI	DARWIN, AUSTRALIA, WEATHER, 0800-2100 UTC
10.563	VOICE		SPY STATION, SPANISH FEMALE VOICE, SENDS 5-DIGIT TRAFFIC, 0400 UTC
10.565	VOICE		SPY STATION, SPANISH FEMALE VOICE, SENDS 5-DIGIT TRAFFIC, 0200 UTC
10.580	VOICE		SPY STATION, ENGLISH FEMALE VOICE, SENDS 5-DIGIT TRAFFIC, 0230 UTC
10.600	VOICE		SPY STATION, SPANISH FEMALE VOICE, SENDS 4-DIGIT TRAFFIC, 0100, 0200, 0230, 2200, 2230, 2300 UTC (WILL ALSO USE 11.532 AT SAME TIME)
10.60330	RTTY		50 BAUD
10.610	AM		RADIO EGYPT
10.6105	RTY	SUA	RY'S, 50 BAUD, BUADOT
10.63630	CW	KKN50	QRA QRA QRA DE KKN50 KKN50 KKN50 QSX 6/10/12/15 K (will TX once per min.)
10.6355	RTY	SUC	RY'S, 50 BAUD, BAUDOT, RXREV OFF
10.640	VOICE		SPY STATION, ENGLISH FEMALE VOICE, SENDS 5-DIGIT TRAFFIC, 0500 UTC
10.665	VOICE		SPY STATION, ENGLISH FEMALE VOICE, SENDS 3-2 DIGIT TRAFFIC, 0600 UTC 4-DIGIT TRAFFIC, 0000,0100, 0200, 0300, 0400, 1400, 1800, 2200, 2300 UTC 5-DIGIT TRAFFIC, 0130, 0200, 0300, 0330, 0400 UTC
10.689	AM		VOICE OF AMERICA
10.710	USB	NAU	USN, CEIBA, PTR
	VOICE		SPY STATION, SPANISH FEMALE VOICE, SENDS 5-DIGIT TRAFFIC, 0500, 0530 UTC
10.713	VOICE		SPY STATION, ENGLISH FEMALE VOICE, SENDS 5-DIGIT TRAFFIC, 0530 UTC
10.715	VOICE		SPY STATION, ENGLISH FEMALE VOICE, SENDS 5-DIGIT TRAFFIC, 0300, 0400, 0430, 0530 UTC

Frequency	Mode	Call Sign	Service / Times
10.720 MHz	VOICE		SPY STATION, SPANISH FEMALE VOICE, SENDS 4-DIGIT TRAFFIC, 0500 UTC ENGLISH FEMALE VOICE 0500, 0530 UTC
10.740	RTTY		U.S. EMBASSY, LONDON, ENGLAND
10.760	AM		VOICE OF AMERICA
10.780	USB		NASA AIR/GROUND
10.840	USB	MKG	R.A.F., LONDON
	VOICE		SPY STATION, SPANISH FEMALE VOICE, SENDS 5-DIGIT TRAFFIC, 0530 UTC
10.857	VOICE		SPY STATION, ENGLISH FEMALE VOICE, SENDS 5-DIGIT TRAFFIC, 0600 UTC
10.860	RTTY		IRAN NEWS AGENCY
10.863	FAX	NAM	NORFOLK, VA, WEATHER
10.866	AM		VOICE OF AMERICA (Used as relay link)
10.869	AM		VOICE OF AMERICA (Used as relay link)
10.877	AM		VOICE OF AMERICA (Used as relay link)
10.880	AM		VOICE OF AMERICA (Used as relay link)
10.9446	CW	CFH	NAWS DE CFH II ZKR F1 3287 4160.6 6238.6 8302 12489 12626.4 KHZ AR
10.9495	CW	FEK1	VVV'S CHL5 CHL5 CHL5 DE FEK1 FEK1 FEK1
10.965	RTTY		JORDAN NEWS AGENCY
10.972	AM		VOICE OF AMERICA (Used as relay link)
10.9446	CW	CFH	NAWS DE CFH II ZKR F1 3287 4160.6 6238.6 8302 12489 12626.4 KHZ AR
10.9495	CW	FEK1	VVV'S CHL5 CHL5 CHL5 DE FEK1 FEK1 FEK1
10.980	FAX	RDD7	MOSCOW, USSR, 0000-2400 UTC
10.990	RTY	CLM29	RY'S TESTING DE CLM29
10.991	VOICE		SPY STATION, GERMAN FEMALE VOICE, SENDS 5-DIGIT TRAFFIC, 0600 UTC

DATE	FREQ.	MODE	TIME	STATION	SIGNAL	COMMENTS	SWL SENT	SWL REC'D

Frequency	Mode	Call Sign	Service / Times
11.0056 MHz	RTTY	NPX	SOUTH POLE STATION, U.S. NAVY, ANTARCTICA
11.011	FAX	JAL	JAPAN WEATHER
11.030	FAX	AXM34	CANBERRA, AUSTRALIA, 0000-2400 UTC
11.070	RTTY		U.S NAVY MARS
11.072	CW	KRH51	U.S. DEPT OF STATE, LONDON GREAT BRITAIN
	CW		AND 50, 75 BAUD RTTY
11.075	ARQ		INTERPOL FREQUENCIES
11.0865	FAX	GHA	BRACKNELL, UNITED KINGDOM, 0000-2400 UTC
11.089	FAX	KVM	NOAA WEATHER
11.090	AM		VOICE OF AMERICA
11.100	USB		PLO NET OF EL FATAH
11.104	USB		NASA MISSION FREQ
11.108	VOICE		SPY STATION, GERMAN FEMALE VOICE, SENDS 5-DIGIT TRAFFIC, 0000, 0030, 0200, 0600, 0630, 1830 UTC
11.275	CW		SPY STATION, MORSE CODE, SENDS 5-DIGIT TRAFFIC, 0930 UTC
11.110	USB		SKYKING BROADCASTS
11.114	CW	CMU	W7TE DE CMU967 (Cuban, Soviet Navy)
11.118	USB		SKYKING BROADCASTS
11.1135	CW	ZTD	CMU967 DE ZTD QSA ? QTC NR ? AS K (Control can be heard on 07692.50)
11.14140	CW	KRH50	QRA QRA QRA DE KRH50 KRH50 KRH50 QSX 5/7/11/13/16/20 IMI QSX 5/7/11/13/16/20 K (Will transmit call once every minute)
11.14735	FAX	LOK	ARGENTINIAN NAVY
11.152	AM		VOICE OF AMERICA
11.165	CW	RCF	MINISTRY OF FOREIGN AFFAIRS, USSR, MOSCOW
	CW		AND 50, 75 BAUD RTTY
11.176	USB		MILITARY AIRLIFT COMMAND
	USB		SKYKING BROADCASTS
11.179	USB		USAF GLOBAL CONTROL AND COMMAND
11.193	USB/CW		RUSSIAN AIRLINE AEROFLOT
11.200	USB		USAF (Used for SAM aircraft)
	USB		INTERNATIONAL WEATHER REPORTS
11.2015	USB	NMN	U.S. COAST GUARD

Frequency	Mode	Call Sign	Service / Times
11.214 MHz	USB		USAF AWACS EARLY WARNING
11.220	USB		SKYKING BROADCASTS
11.226	USB		X-RAY 905
11.230	CW	CLP1	MINISTRY OF FOREIGN AFFAIRS, HAVANA, CUBA. CW AND RTTY 50 BAUD
	RTTY		NORTH KOREAN NEWS AGENCY
11.233	USB		U.S. AIRFORCE
11.239	USB		McCLELLAN AIR FORCE BASE
	USB		SKYKING BROADCASTS
11.241	ARQ		EGYPTAIN EMBASSY IN WASHINGTON, D.C.
11.243	USB	U7E	U.S. AIR FORCE (C/S changes daily)
	USB		SIERRA 393
	USB		SKYKING BROADCASTS
11.246	USB		McDILL AIR FORCE BASE
	USB		SKYKING BROADCASTS
11.252	USB		NASA MISSION FREQ
11.267	USB	U7E	U.S. AIR FORCE (C/S changes daily)
	USB		SKYKING BROADCASTS
11.275	CW		SPY STATION, MORSE CODE, SENDS 5-DIGIT TRAFFIC, 0930 UTC
11.282	USB		INTERNATIONAL AIRLINES
11.288	USB		U.S. AIR FORCE
11.297	USB		RUSSIAN WEATHER REPORTING
11.312	CW/USB		RUSSIAN AIRLINE AEROFLOT
11.330	AM		RADIO PEKING, 0600-1000 UTC
11.335	AM		NORTH KOREA RADIO, 0000 UTC
11.339	USB		U.S. NAVY
11.342	USB		COMMERICAL AIRCRAFT
	USB		DISPATCH UNITED AIR LINES
11.348	CW		AEROFLOT (commerical flights)
11.359	CW		SPY STATION, MORSE CODE, SENDS 5-DIGIT TRAFFIC, 0500 UTC
11.368	USB		NATIONAL HURRICANE SERVICE
11.380	VOICE		SPY STATION, SPANISH FEMALE VOICE, SENDS 4-DIGIT TRAFFIC 1900, 1930 UTC
11.387	USB	ATTF	AIR LINE WEATHER FROM AUSTRALIA
11.406	CW	COY	DE COY (Cuban, sending 5-digit letters)
11.407	USB		NASA MISSION FREQ
11.408	USB		SKYKING BROADCASTS
11.441	USB		USAF NORAD HQ

11

Frequency	Mode	Call Sign	Service / Times
11.457 MHz	CW	KKN50	QRA QRA QRA DE KKN50 KKN50 KKN50 QSX 6/10/11/15 K (WILL TRANSMIT CALL ONCE EVERY MINUTE)
11.460	AM		KOREAN BROADCASTS SERVICE, 0245-0300 UTC
11.4755	CW/FSK	UHF3	DE UHF3 (Sends High Speed Morse and RTTY)
	FAX	HMY	NORTH KOREA
11.490	VOICE		SPY STATION, SPANISH FEMALE VOICE, SENDS 4-DIGIT TRAFFIC, 0100, 0130 UTC
11.492	VOICE		SPY STATION, SPANISH FEMALE VOICE, SENDS 4-DIGIT TRAFFIC, 1830 UTC
11.494	USB		SKYKING BROADCASTS
	USB		LOOKING GLASS, SIERRA 311
11.497	RTTY		KUWAIT NEWS SERVICE
11.532	VOICE		SPY STATION, SPANISH FEMALE VOICE, 4-DIGIT TRAFFIC, 0100, 0130, 0200, 0230, 0300, 0400, 0600, 2200, 2300, 2330 UTC (WILL ALSO USE 10.600, 5.930, 5.812)
11.538	ARQ		INTERPOL FREQUENCIES
11.545	VOICE		SPY STATION, SPANISH FEMALE VOICE, SENDS COMBO TRAFFIC, 0400, 0430, 0450, 0600 UTC
11.548	USB		NASA MISSION FREQ
11.550	AM	WYFR	FAMILY RADIO NETWORK, (English Service to India), 1302-1502 UTC
	VOICE		SPY STATION, ENGLISH FEMALE VOICE, SENDS 3-2 DIGIT TRAFFIC, 1400, 1500 UTC
11.565	VOICE		SPY STATION, SPANISH FEMALE VOICE, SENDS 5-DIGIT TRAFFIC, 0230, 0400 UTC
11.580	AM		VOICE OF AMERICA, (English to American Republics service), 0000-0230 UTC (Spanish to American Republics), 0930-1130 UTC

Frequency	Mode	Call Sign	Service / Times
11.580 MHz	AM		CHRISTIAN SCIENCE MONITOR, 1600-1655 UTC (Saturday & Sunday), 1605-1755 UTC
	AM		VOICE OF FREE CHINA, (English Service),(Europe Beaming, can be heard in North America), 2200-2300 UTC
11.585	AM		THE VOICE OF ISRAEL, (English Service), 1100-1130, 1815-1830 UTC
11.588	AM		KOL ISRAEL ENGLISH, 0400-0430, 1000-1030 UTC
11.600	RTTY	CLN	HANANA, CUBA
	AM		RADIO BEIJING, CHINA, (English Service), 1200-1400 UTC
11.604	AM		KOL ISRAEL RUSSIAN SECTION, 0500-0540, 1600-1655, 1900-1940 2100-2155, 2300-2355 UTC
	AM		KOL ISRAEL ENGLISH SECTION, 2300-2330, 0000-0025, 0100-0125, 0400-0430 UTC
11.605	AM		THE VOICE OF ISRAEL, (English Service), 0000-0030, 0100-0130, 0200-0230, 0500-0515, 2000 UTC
	VOICE		SPY STATION, ENGLISH FEMALE VOICE, SENDS 5-DIGIT TRAFFIC, 0200 UTC
11.606	CW	CFG	CO4 DE CFG, (Possible Cuban, will make cw call up than pass
	RTTY		traffic/can not hear outstation) 0400 UTC
11.607	USB		SKYKING BROADCASTS
11.608	CW	OVG	VVV VVV VVV OVG 8/12/16 ZKR 6 7 11 18 MC/S
11.620	AM		RADIO INDIA, 2000-2200 UTC
	CW	HLL	CQ CQ CQ DE HLL3
11.625	AM		RADIO FRANCE, 0200-0400 UTC
	VOICE		SPY STATION, ENGLISH FEMALE VOICE, SENDS 3-2 DIGIT TRAFFIC, 1300, 1330, 1400 UTC

Frequency	Mode	Call Sign	Service / Times
11.628 MHz	CW	OVG	VVV VVV VVV OVG 2/12/16 ZKR 6 8 11 18 MC/S
11.632	VOICE		SPY STATION, SPANISH FEMALE VOICE, SENDS 5-DIGIT TRAFFIC, 0500 UTC
	CW		SPY STATION, MORSE CODE, SENDS 5-DIGIT TRAFFIC, 0130, 0200, 1000, 2230, 2300 2330 2345 UTC (WILL USE 12.214 ON HALF HOUR)
11.635	CW	KRH51	U.S. DEPT OF STATE, LONDON, GREAT BRITAIN
			CW AND 50, 75 BAUD RTTY
11.6425	ASCII		100 BAUD ASCII
11.645	AM		VOICE OF GREECE, 0130-0200 UTC
11.655	AM		THE VOICE OF ISRAEL, (English Service), 1815-1830, 2230-2300 UTC
	AM		RADIO BEIJING, CHINA, 0000-0230 UTC
	AM		KOL ISRAEL RUSSIAN SECTION, 0500-0540 UTC
11.650	RTTY/CW		VIET NAM EMBASSY TO YEMEN
11.660	AM		RADIO NETHERLANDS INTERNATIONAL, (English Service), 2000-2200 UTC, (Spanish Service), 1130-1155 UTC
	VOICE		SPY STATION, SPANISH FEMALE VOICE, SENDS 4-DIGIT TRAFFIC, 0330 UTC
11.663	CW		SPY STATION, MORSE CODE, SENDS 5-DIGIT TRAFFIC, 0200, 1000 UTC (WILL ALSO USE 11.632)
11.667	CW		SPY STATION, MORSE CODE, SENDS 5-DIGIT TRAFFIC, 0900, 0930, 0945 UTC
11.675	AM		KOL ISRAEL RUSSIAN SECTION, 0735-0800, 1200-1230 UTC
	AM		RADIO MOSCOW, (East Coast), 0400-0500 UTC, (Sept thru Nov)
	AM		RADIO MOSCOW WORLD SERVICE, (English Service) 0300-0500 UTC

Frequency	Mode	Call Sign	Service / Times
11.680 MHz	AM		THE VOICE OF BAGHDAD, (English Service),(Baghdad Betty), 1000-1200, 1600-1800, 2000-2200 UTC
	AM		VOICE OF FREE CHINA, (English Service), 0200-0300 UTC
	AM		VOICE OF AMERICA, (Service to Brazil), 2300-2400 UTC
11.685	AM		RADIO NORWAY INTERNATIONAL, 0300-0500 UTC
11.690	AM		RED CROSS BROADCASTING SERVICE, (English), 1310-1327 UTC
	AM		SWISS RADIO INTERNATIONAL, 1330-1400 UTC
11.691	VOICE		SPY STATION, SPANISH FEMALE VOICE, SENDS 4-DIGIT TRAFFIC, 1830 UTC
11.695	AM		VOICE OF AMERICA (English to Caribbean service) 0000-0100 UTC (Latin America), 2130-2200 UTC
	AM		RADIO BEIJING, CHINA, (English Service), 0400-0500 UTC
11.700	AM		RADIO NORTH KOREA, 0000-0245 UTC
11.705	AM		RADIO GERMANY, (English Service), 0500 UTC
	AM		VOICE OF AMERICA, (English to South Asia), 0100-0300 UTC (Central Asia), 0000-0100 UTC
	AM		CHRISTIAN SCIENCE WORLD SERVICE, 0600-0855 UTC
11.710	AM		VOICE OF AMERICA, (Russian to USSR), 1700-1900 UTC

Frequency	Mode	Call Sign	Service / Times
11.710 MHz	AM		RADIO MOSCOW, (English Service, East Coast), 2300-0400 UTC (May 5 - Aug 31)
	AM		RADIO MOSCOW WORLD SERVICE, (English Service) 0000-0400, 2200-0000 UTC
	AM		RADIO ARGENTINA, 0100-0400 UTC
11.715	AM		RADIO BEIJING, CHINA, (English Service), 0000-0100, 0300-0400 UTC
	AM		VOICE OF AMERICA, (English to Pacific service) 1200-1330 UTC
	AM		RADIO SOUTH KOREA, (KBS), (Spanish Service), 0130-0230 UTC
	AM	WYFR	FAMILY RADIO NETWORK, (Spanish Service), 0100-0300 UTC
	AM		RADIO NETHERLANDS INTERNATIONAL, (Spanish Service), 2330-0025 UTC
11.720	AM		VOICE OF AMERICA, (English to Pacific service) 1000-1200 UTC
	AM		RADIO AUSTRALIA, 1130-1530 UTC
	AM		RADIO NETHERLANDS INTERNATIONAL, 0030-0425 UTC
11.725	AM		RADIO FREE EUROPE, (Romanian), 1200-1800 UTC
	AM		RADIO SOUTH KOREA, (KBS), (Spanish Service), 1015-1100 UTC
	AM	WYFR	FAMILY RADIO NETWORK, (Spanish Service), 1100-1400 UTC
11.730	AM		RADIO MOSCOW, (English Service, East Coast), 2330-0300 UTC, (Aug 1 - Sept 28)

Frequency	Mode	Call Sign	Service / Times
11.730 MHz	AM	HCJB	THE VOICE OF THE ANDES (English Service) 0700-0830 UTC
11.735	AM		VOICE OF AMERICA, (English to Middle East/Europe service), 0800-1300 UTC (Vietnamese to East Asia), 1230-1330 UTC (English to VOA Europe), 0800-1000 UTC
	AM		RADIO YUGOSLAVIA, 0000 UTC
	AM		RADIO JAPAN, 2300-0000 UTC
	AM		RADIO MOSCOW, (East Coast), 0000-0400, (March thru March)
	AM		RADIO FINLAND, (English Service) 1230-1330 UTC
11.740	AM		VOICE OF FREE CHINA, TAIWAN, (Spanish Service), 0200-0400 UTC
	AM		VOICE OF AMERICA, (Russian to USSR) 1600-1800 UTC (Middle East), 0800-1100 UTC
	AM		VOICE OF FREE CHINA, (English Service), 0300-0400 UTC
	AM		KOREAN BROADCASTS SERVICE, 1115-1130, 1415-1430 UTC
11.750	AM		VOICE OF AMERICA, (English to Middle East/Europe service) 0500-1000 UTC
	AM		RADIO MOSCOW, (English Service, East Coast) 2300-0300 UTC, (Sept thru sept 28)
	AM		BRITISH BROADCASTING CORPORATION, 2200-0330, 1100-1700 UTC
11.755	AM		RADIO FINLAND, 0600-0700 UTC

Frequency	Mode	Call Sign	Service / Times
11.755 MHz	AM		RADIO BEIJING, CHINA, (English Service), 0900-1000 UTC
	AM		RADIO FINLAND (English Service) 0245-0345 UTC
11.760	AM		BRITISH BROADCASTING CORPORATION, 0330-1330 UTC
	AM		VOICE OF AMERICA, (English to Pacific service) 2200-0100 UTC
	AM		RADIO CUBA, (English Service) 0600-0800 UTC,
11.765	AM		RADIO DEUTSCHE WELLE, 0400-0500 UTC
11.770	AM		RADIO FREE EUROPE, (Romanian), 1200-1800 UTC
	AM		RADIO FREE AFGHANISTAN, 0230-0330 UTC
	AM		RADIO NORWAY INTERNATIONAL, 1200-1300 UTC
11.775	AM		BRITISH BROADCASTING CORPORATION, WORLD SERVICE, (English Service), 1500-1730 UTC, (Spanish Service) 0000-0200, 0300-0430 UTC
11.780	AM		RADIO MOSCOW, (English Service, East Coast) 0000-0400 UTC, (May 5 - Aug 31)
	AM		RADIO MOSCOW WORLD SERVICE, (English Service) 0000-0300, 2100-2200 UTC
	AM		VOICE OF AMERICA, (Urdu to South Asia), 0100-0130 UTC
	AM		RADIO AUSTRIA INTERNATIONAL, 1530-1600, 1130-1200, 1330-1400 UTC

Frequency	Mode	Call Sign	Service / Times
11.785 MHz	SSB		RADIO CUBA, 0300-0600 UTC, BEAMING TOWARDS SOUTH AMERICA
	AM		VOICE OF AMERICA, (Vietnamese to East Asia), 2230-2330 UTC
11.790	AM		ISLAMIC REPUBLIC OF IRAN BROADCASTING, (English Service), (East Asia), 1130-1230 UTC
11.795	AM		RADIO MOSCOW, (Germany Service), 0500 UTC
11.800	AM		RADIO AUSTRALIA, 0830-1100 UTC
	AM		RADIO ITALY, 0100-0120, 1935-1955 UTC
11.805	AM		VOICE OF AMERICA, (English to Middle East/Europe service) 0600-0700, (Russian to USSR), 1600-1700, 1800-1900 UTC, (Urdu to South Asia), 1330-1430 UTC
	AM		RADIO SOUTH KOREA, (KBS), (Spanish Service), 0130-0230 UTC
11.810	AM		THE VOICE OF BAGHDAD, (English Service), 0230-0430 UTC
	AM		RADIO DEUTSCHE WELLE, 0300-0400 UTC
	AM		RADIO CHILE, 0200-0300 UTC
	AM		RADIO SOUTH KOREA, (KBS), (English Service), 0600-0700 UTC
11.815	AM		RADIO FREE EUROPE, (Romanian), 0400-1200 UTC, (Polish), 1500-1800 UTC, (Romanian), 2000-2200 UTC

Frequency	Mode	Call Sign	Service / Times
11.815 MHz	AM		RADIO JAPAN, 1100-1200, 1700-1800, 2100-2200, 2300-0000 UTC (Beaming Asia)
	AM		POLISH RADIO WARSAW, (English Service), 1300-1355 UTC
11.820	AM		RADIO HAVANA, CUBA, 0000-0400 UTC
	AM		BRITISH BROADCASTING COPRORATION, WORLD SERVICE, (English Service), 1300-1500 UTC, (Spanish Service), 0000-0200, 0300-0430 UTC
11.825	AM		VOICE OF AMERICA, (Russian to USSR), 1800-2300 UTC, (Ukrianian to the USSR), 1600-1800 UTC
	AM		RADIO FREE EUROPE, (Pashto) 0230-0300 UTC, (Dari) 0300-0330
11.830	AM		THE VOICE OF BAGHDAD, (English Service), 0230-0430, 1700-1900 UTC
	AM		VOICE OF AMERICA, (Service to Latin America), 2330-0030 UTC
	AM	WFYR	FAMILY RADIO NETWORK, 1200-1700 UTC, (Spanish Service) 1000-1100 UTC
	AM		ORGANIZATION OF AMERICAN STATES, (All broadcasts in English except for Spanish), @ 2345-0034 UTC
	AM		UNITED NATIONS RADIO, (Spanish Service), 0015-0020 UTC

Frequency	Mode	Call Sign	Service / Times
11.835 MHz	AM		VOICE OF AMERICA, (French Service) 0400-0500 UTC, (English to Africa Service) 0300-0700 UTC, (Vietnamese to East Asia) 1230-1330 UTC, (Russian to USSR) 1200-1400 UTC
	AM		RADIO JAPAN, 2300-0000 UT
	AM		RADIO NETHERLANDS INTERNATIONAL, 0030-0125 UTC
	AM		UNITED NATIONS RADIO, (Spanish Service), 1930-2000 UTC
11.840	AM		RADIO MOSCOW, 1400-2200 UTC
	AM		RADIO MOSCOW WORLD SERVICE, (English Service) 1000-2200 UTC
	AM		RADIO JAPAN, 0100-0200, 0900-1000, 1100-1200 UTC (Asia)
	AM		RADIO BEIJING, CHINA, (English Service), 0500-0600 UTC
11.850	AM		RADIO MOSCOW, (English Service, East Coast), 0000-0400 UTC (May 5 - Aug 31)
	AM		RADIO MOSCOW WORLD SERVICE, (English Service) 0000-0500 UTC
	AM		RADIO JAPAN 1900-1930 UTC (Oceania)
11.855	AM		RADIO FREE EUROPE, (Uzbek), 0100-0300 UTC, (Tatar-Bashkir) 0300-0600 UTC, (Russian), 0600-0900 UTC

Frequency	Mode	Call Sign	Service / Times
11.855 MHz	AM	WYFR	FAMILY RADIO NETWORK, 0100-0545 UTC, (Spanish Service), 0200-0600 UTC
	AM		RADIO CANADA INTERNATIONAL (Mon thru Fri) 1200-1400 UTC
11.860	AM		THE VOICE OF BAGHDAD, (English Service), 0200-0300,1700-1900, 2100-2300 UTC
	AM		VOICE OF FREE CHINA, TAIWAN, 0200-0400 UTC
	AM		RADIO NORWAY INTERNATIONAL, 1300 UTC
	AM		BRITISH BROADCASTING CORP., 0800-0930 UTC
	AM		RADIO MOSCOW, (East Coast), 0400-0500 UTC, (Mar thru Mar)
11.865	AM		VOICE OF AMERICA, (Service to Europe), 1430-1700 UTC
	AM		RADIO NORWAY INTERNATIONAL, 0400 UTC
	AM		RADIO JAPAN, 1400-1600, 1700-1800, 1900-1930 UTC
11.870	AM		VOICE OF AMERICA, (English to Pacific Service) 1900-2400 UTC
	AM		RADIO NORWAY INTERNATIONAL, 1500-1600 UTC
11.875	AM		RADIO FREE EUROPE, (Azerbaijan), 0000-0100 UTC, (Armenian) 0100-0200 UTC, (Georgian) 0200-0300 UTC, 1800-2000 UTC (Armenian) 0300-0400 UTC, 1200-1400 UTC,

Frequency	Mode	Call Sign	Service / Times
11.875 MHz	AM		(Azerbaijan) 0400-0500 UTC, (Tatar-Bashkir) 0900-1000 UTC, (Azerbaijan) 1100-1200 UTC, (Azerbaijan) 1400-1600 UTC,
	AM	WYFR	FAMILY RADIO NETWORK, (Spanish to Europe), 2200-2300 UTC
11.880	AM		RADIO AUSTRALIA, 2100-0800 UTC
	AM		VOICE OF AMERICA, (Service to South Asia), 1600-1700 UTC, (Africa), 1800-1900 UTC
	AM		SPANISH NATIONAL RADIO, 0000 UTC
11.885	AM		RADIO FREE EUROPE, (Russian) 0000-2400 UTC
11.890	AM		VOICE OF AMERICA, (Spanish to American Republics) 1200-2200 UTC, (Africa), 0430-0700 UTC
	AM		RADIO DEUTSCHE WELLE, 0300-0600 UTC
11.895	AM		RADIO FREE EUROPE, (Romanian), 1400-1600 UTC, (Hungarian), 1600-2200 UTC
11.905	AM		VOICE OF AMERICA, (English to Middle East/Europe service) 0000-0330, 1330-1700, 1800-2400 UTC (English to VOA Europe) 0300-0330 UTC

Frequency	Mode	Call Sign	Service / Times
11.905 MHz	AM		RADIO FREE EUROPE, (Kazak) 1100-1400 UTC, (Kirghiz) 1400-1500 (Tatar-Bashkir) 1600-1700 UTC, (Uzbek) 1700-1800 UTC
	AM		RADIO CANADA INTERNATIONAL 2200-2300 UTC
11.910	AM		RADIO AUSTRALIA, 1600-2030 UTC
	AM	HCJB	THE VOICE OF THE ANDES (Spanish Service) 1200-1430 UTC
11.915	AM		RADIO FREE EUROPE, (Kazak), 0000-0200 ,2300-2400 UTC (Russian), 0200-0600 UTC,
	AM	WYFR	FAMILY RADIO NETWORK, 2300-2400 UTC, (English to Europe), 0500-0600 UTC
11.920	AM		RADIO MOSCOW WORLD SERVICE, 1700 UTC
	AM		VOICE OF AMERICA, (English to Africa service) 1600-2000 UTC
11.925	AM		RADIO FREE EUROPE, (Romanian), 1600-2000 UTC
	AM		RADIO NORWAY INTERNATIONAL, 2300-0100 UTC
	AM	HCJB	THE VOICE OF THE ANDES (English Service) 1130-1600, 0500-0700, 0730-1130 UTC
11.930	AM		RADIO AUSTRALIA, 2000-0800 UTC

Frequency	Mode	Call Sign	Service / Times
11.930 MHz	AM		VOICE OF AMERICA, (Spanish to Cuba), 1400-2300 UTC (Russian to USSR) 0800-1100 UTC
	AM		CBS-TRANS-WORLD RADIO, (Christain Radio), 0300-0400 UTC
	AM		ISLAMIC REPUBLIC OF IRAN BROADCASTING, (English Service), (East Asia), 1130-1230 UTC
11.935	AM		VOICE OF AMERICA, (Spanish to American Republics), 0930-1130 UTC
	AM		RADIO FREE EUROPE, (Tajik) 0100-0200 UTC, (Ukrainian) 0200-0600 UTC, 1700-2300 UTC
11.938	AM		RADIO CAMBODIA, 1200 UTC
11.940	AM		VOICE OF GREECE, 0430 UTC
	AM		BRITISH BROADCASTING CORPORATION, 0100-1300 UTC
	CW		SPY STATION, MORSE CODE, SENDS 5-DIGIT TRAFFIC, 0500 UTC
11.945	AM		THE VOICE OF THE UNITED EMIRATES, (English Service)
	AM		VOICE OF AMERICA, (Spanish to American Republics) 1200-1500 UTC, (Uzbek to the USSR), 0000-0100 UTC
	AM		BRITISH BROADCASTING COPR., 2200-0030 UTC
	AM		RADIO EXTERIOR DE ESPANA (Spanish Service) 0900-1900 UTC

Frequency	Mode	Call Sign	Service / Times
11.950 MHz	AM		RADIO MOSCOW, (East Coast), 0000-0500 UTC, (Mar thru Mar)
	AM		RADIO CUBA, 0100-0600 UTC, (English service)
11.955	AM		BRITISH BROADCASTING CORPORATION, 0300-1300 UTC
	AM		VOICE OF AMERICA, (Service to East Asia), 2130-2200 UTC
	AM		BRITISH BROADCASTING COPR., 2200-0030 UTC
	AM		SWISS RADIO INTERNATIONAL, 1830-1900 UTC
	AM		RADIO CANADA INTERNATIONAL 1300-1600 UTC
11.960	AM		RADIO NEW ZEALAND, (English Service), 1300-1330 UTC
	AM		RADIO FREE EUROPE, (Russian), 0400-0600 UTC
	AM		VOICE OF AMERICA, (Turkish to Europe/Middle East), 2000-2100
	AM	HCJB	THE VOICE OF THE ANDES (Spanish Service) 1030-1300, 2030-0500 UTC
11.965	AM		VOICE OF AMERICA (Russian to USSR), 0800-1000 UTC, (China), 1000-1500 UTC

Frequency	Mode	Call Sign	Service / Times
11.970 MHz	AM		RADIO FREE EUROPE, (Estonian), 0400-0430 UTC, (Latvian), 0430-0500 UTC, (Lithuanian), 0500-0530 ,2100-2200 UTC, (Latvian), 1500-1600 UTC, (Lithuanian), 1600-1700 UTC, (Estonian), 1800-1900 UTC, (Latvian), 2000-2100 UTC,
11.975	AM		RADIO CAIRO, (United Nations Radio Re-broadcasts),(Monday) 1650-1705 UTC
11.980	AM		RADIO MOSCOW, (English Service, East Coast) 0200-0400 UTC, (May 5 - Sept 28)
	AM		RADIO MOSCOW WORLD SERVICE, (English Service) 0400-0500 UTC
	CW		SPY STATION, MORSE CODE, SENDS 5-DIGIT TRAFFIC 0200 UTC
11.990	AM		RADIO BAGHDAD, 2100-2200 UTC
	AM		RADIO CZECHOSLOVAKIA, (English Service), 0000-0027 UTC
11.995	AM		RADIO MOSCOW, 0200-0500 UTC
	AM		RADIO MOSCOW WORLD SERVICE, (English Service) 1300-1800 UTC
	AM		BRITISH BROADCASTING CORP., 2300-0030 UTC

Frequency	Mode	Call Sign	Service / Times
12.000 MHz	AM		RADIO MOSCOW, (English Service, West Coast) 0530-0800 UTC, (Sept 1 thru Sept 28)
	AM		RADIO AUSTRALIA, 1430-2100 UTC
12.010	AM		RADIO MOSCOW, (English Service, West Coast) 0530-0800 UTC, (Sept 1 thru Sept 28), 0330-0500 UTC, (Sept thru Sept)
	AM		RADIO AUSTRIA INTERNATIONAL, 1830-1900 UTC
12.020	AM		RADIO "THE VOICE OF VIETNAM", (English Service), 1000-1030, 1230-1300, 1330-1400, 1600-1630, 1800-1830, 1900-1930, 2030-2100, 2330-2400 UTC, (Spanish Service), 1100-1130, 2000-2030 UTC
12.02230	CW	KKN50	QRA QRA QRA DE KKN50 KKN50 QSX 6/10/12/16 K (Will transmit call once every minute)
12.025	FAX	PWZ	WEATHER, RIO de JANEIRO, BRAZIL
12.030	AM		RADIO MOSCOW, (English Service, West Coast) 0430-0800 UTC (Sept thru Sept 28)
	AM		RADIO MOSCOW WORLD SERVICE, (English Service) 1500-1700 UTC
12.035	AM		SWISS RADIO INTERNATIONAL, (English Service), 2200-2230, 0000-0030, 0200-0230, 0400-0430 UTC, (Spanish Service), 2230-2300 UTC
	AM		RED CROSS BROADCASTING SERVICE, (English), 0310-0327 UTC
	AM		RADIO EXTERIOR DE ESPANA (Spanish Service) 0900-1900 UTC

Frequency	Mode	Call Sign	Service / Times
12.040 MHz	AM		RADIO MOSCOW, (English Service, East Coast) 2300-0300 UTC, (May 5 - July 31)
	AM		RADIO MOSCOW WORLD SERVICE, (English Service) 0000-0500, 2100-2200 UTC
12.050	AM		RADIO MOSCOW, (English Service, East Coast) 2300-0400 UTC, (May 5 - Sept 28)
	AM		RADIO MOSCOW WORLD SERVICE, (English Service) 0000-0300, 1300-0000 UTC
12.056	CW	RIW	DE RIW, (High speed traffic and RTTY) (Russian Navy)
12.065	RTTY		IRAN NEWS AGENCY
	AM		KOL ISRAEL RUSSIAN SECTION, 2100-2155 UTC
12.070	USB		LOOKING GLASS
	RTTY		JAE KYODO PRESS, TOKYO NEWS AGENCY, (JAE58/JAT28)
	AM		RADIO MOSCOW WORLD SERVICE, (English Service) 1900-2200 UTC
12.076	RTTY		USSR EMBASSY IN WASHINGTON, D.C.
12.079	RTTY		KUWAIT NEWS SERVICE
12.085	AM		RADIO DAMASCUS, SYRIA, 2100 UTC
	VOICE		SPY STATION, SPANISH FEMALE VOICE, SENDS 5-DIGIT TRAFFIC, 0100, 0200, 0300 UTC SENDS 4-DIGIT TRAFFIC 2200 UTC
12.095	AM		BRITISH BROADCASTING CORPORATION, 0000-2400 UTC
12.100	VOICE		SPY STATION, SPANISH FEMALE VOICE, SENDS 5-DIGIT TRAFFIC, 1900 UTC
12.107	USB		NASA MISSION FREQ
12.108	FAX	BAF	BEIJING, PEOPLES REPUBLIC OF CHINA, WEATHER
12.110	VOICE		SPY STATION, SPANISH FEMALE VOICE, SENDS 3-2 DIGIT TRAFFIC, 1400 UTC
12.111	CW	KKN50	QRA QRA QRA DE KKN50 KKN50 KKN50 QSX 6/10/12/16 K (Will transmit call once every minute)

12

Frequency	Mode	Call Sign	Service / Times
12.123 MHz	CW		NAVY, NEW ZEALAND
12.13440	CW	NMN	CQ CQ CQ DE NMN/NAM/NRK/NAR/GXS/AOK QRU K
12.156	VOICE		SPY STATION, SPANISH FEMALE VOICE, SENDS 4-DIGIT TRAFFIC, 0200, 0205, 0230, 0300 UTC (WILL ALSO USE 9.222 AT SAME TIME)
12.170	FAX	RNR4	U.S.S.R.
12.1745	RTTY		NORTH KOREAN NEWS AGENCY
12.180	VOICE		SPY STATION, SPANISH FEMALE VOICE, SENDS 5-DIGIT TRAFFIC, 0530, 0605 UTC
12.185	VOICE		SPY STATION, ENGLISH FEMALE VOICE, SENDS 3-2 DIGIT TRAFFIC, 1400 UTC
12.210	AM		VOICE OF AMERICA
	CW	KWL90	QRA QRA QRA DE KWL90 KWL90 KWL90 (Will transmit call once every minute)\
12.214	CW		SPY STATION, MORSE CODE, SENDS 5-DIGIT TRAFFIC, 0100, 2300 UTC (WILL REPEAT ON 11.362 AT HALF HOUR)
12.219	CW		SPY STATION, MORSE CODE, SENDS 4-DIGIT TRAFFIC, 1200, 1300 UTC
12.222	VOICE		SPY STATION, ENGLISH FEMALE VOICE, SENDS 3-2 DIGIT TRAFFIC, 1200 UTC
12.224	ARQ		INTERPOL FREQUENCIES
12.2385	CW	ONN34	EMBASSY OF REPUBLIC OF IRAN, BRUSSELS, BELGUIM
12.242	USB	JU4Q	NOJ DE JU4Q "LISTENING ON 13080 AND 6501 FOR YOU"
12.245	USB		SHIP TO SHORE PHONE SERVICE
12.260	USB		SHIP TO SHORE PHONE SERVICE
12.263	USB		INTERNATIONAL SSB RADIO TELEPHONY (Receive on 13110.00)
12.275	RTTY	JAE	KYODO PRESS, TOKYO NEWS AGENCY, (JAE58/JAT28)
12.277	USB		NASA MISSION FREQ
12.282	CW		SPY STATION, MORSE CODE, SENDS 3-DIGIT TRAFFIC, 0545, 0555 UTC
12.290	USB		SHIP TO SHORE RADIO
	VOICE		SPY STATION, SPANISH FEMALE VOICE, 5-DIGIT TRAFFIC, 0500, 0530 UTC

Frequency	Mode	Call Sign	Service / Times
12.300 MHz	VOICE		SPY STATION, SPANISH FEMALE VOICE, SENDS 4-DIGIT TRAFFIC, 0100, 0130 UTC
12.319	VOICE		SPY STATION, ENGLISH FEMALE VOICE, SENDS 3-2 DIGIT TRAFFIC, 1500 UTC
	CW		SPY STATION, MORSE CODE, SENDS 5-DIGIT TRAFFIC, 0730 UTC
12.320	FAX	RY076	DE RY076, (Fax in U.S.S.R.)
12.326	USB		INTERNATIONAL SSB RADIO TELEPHONY, (Receive on 13173.00)
12.32875	CW	OVG	VVV VVV VVV OVG/12/19 ZKR 6 7 11 16 MC/S
12.332	USB		INTERNATIONAL SSB RADIO TELEPHONY, (Receive on 13179.00)
12.362	USB		SHIP TO SHIP COMMUNICATIONS
12.367	USB		SHIP TO SHIP COMMUNICATIONS, (Used by Japan)
12.37590	USB		JAPAN FISHING BOATS
12.3915	USB		SKYKING BROADCASTS
12.392	SSB		MARITIME SAFETY AND DISTRESS
12.400	CW	EWZA	NA16 DE EWZA, (Sends 4-digit coded traffic, out station not on same frequency)
12.410	VOICE		SPY STATION, SPANISH FEMALE VOICE, SENDS 5-DIGIT TRAFFIC, 0700 UTC (WILL REPEAT TRAFFIC TWICE)
12.424	CW	9HRG2	WLO WLO WLO DE 9HRG2 9HRG2 GM QTC 1 QSA NIL KKK
12.42920	USB		SHIP TO SHIP, SHIP TO SHORE COMMUNICATIONS.
12.43230	USB		SHIP TO SHIP, SHIP TO SHORE COMMUNICATIONS.
12.4355	USB		SHIP TO SHIP, SHIP TO SHORE COMMUNICATIONS.
12.444	CW	HCJG	VVV VVV DE HCJG HCJG HCJG QRA QTC ? K
12.445	VOICE		SPY STATION, GERMAN FEMALE VOICE, SENDS 5-DIGIT TRAFFIC, 0800 UTC
12.450	VOICE		SPY STATION, SPANISH FEMALE VOICE, SENDS 4-DIGIT TRAFFIC, 2330 UTC
12.460	CW		SPY STATION, MORSE CODE, SENDS 5-DIGIT TRAFFIC, 0600 UTC

Frequency	Mode	Call Sign	Service / Times
12.470 MHz	VOICE		SPY STATION, ENGLISH FEMALE VOICE, SENDS 5-DIGIT TRAFFIC, 0100, 0130, 0200 UTC
12.471	CW	URF1	VKD VKD DE URF1 URF1 QRU ?
12.4765	CW	PA4O	DE PA4O PA4O MSG NW ?
12.49320	ALIST		100 BAUD ALIST RXREV OFF
12.521	USB		U.S. COAST GUARD NEW ORLEANS, LA
12.5227	RTTY		51 BAUD 6-BIT RTTY RXREV ON\
12.540	VOICE		SPY STATION, SPANISH FEMALE VOICE, SENDS 5-DIGIT TRAFFIC, 0300 UTC
12.54540	USB		JAPANESE MARITIME AGENCY
12.54660	USB		JAPANESE MARITIME AGENCY
12.5475	CW	ROP5	DZK DZK DZK DE ROP5 ROP5 PSE QRJ QTC ? K
12.54840	CW	SYOM	UFH DE SYOM SYOM SYOM QRA ? QTC K
12.5525	CW	HL4J	HEBF DE HL4J HL4J QSW 452
	CW	SQGP	UQK UQK UQK DE SQGP SQGP QSS 450 QSS 450 K K
	CW	UUAF	LMRK LMRK DE UUAF UUAF UUAF UUAF QSW ? ? ? K
	CW	HGHI	5LG 5LG DE HGHI HGHI QSA ? QTC QTC ? K
	CW	COJ	CLH CLH CLH DE COJ COJ QTC 5 K
12.558	CW	EWI	DE EWI EWI EWI
12.580	ARQ	GKE5	
12.5815	ARQ	WLO	(Receive on 12479.00)
	USB		U.S. COAST GUARD GIVES WEATHER UP-DATES EVERY HALF-HOUR, THIS ANNOUNCEMENT IS DONE IN COMPUTER ELECTRONIC VOICE.
12.5835	ARQ	CBV	
12.584	ARQ	VIS	
12.5845	ARQ	WLO	(Receive on 12482.00)
12.585	ARQ	HPP/MTX	
12.5855	ARQ	PCH	
12.586	ARQ	VIS	
	ARQ	KPH	
12.5865	ARQ	WLO	(Receive on 12484.00)
	ARQ	NMO	
12.589	ARQ	WCC	
	ARQ	NMO	
12.5895	ARQ	NMN	
12.5905	ARQ	KPH	
12.591	ARQ	KLB	

Frequency	Mode	Call Sign	Service / Times
12.5915MHz	ARQ	WLO	(Receive on 12489.00)
12.5925	ARQ	NMN	
	ARQ	OXZ	
12.5935	ARQ	WLO	(Receive on 12491.00)
12.594	ARQ	GKP5	
12.5945	ARQ	PPR	
12.595	ARQ	PCH	
	ARQ	WCC	
12.5955	ARQ	SPB	
12.596	ARQ	WLO	(Receive on 12493.50)
	ARQ	GCH	
12.597	ARQ	SPC	
	ARQ	PCH	
12.5975	ARQ	SPB	
	ARQ	UAH	
12.598	ARQ	WCC	
12.5985	ARQ	SPA	
12.599	ARQ	WLO	(Receive on 12496.50)
12.6005	ARQ	KPH	
12.601	ARQ	ZSC	DE ZSC SITOR SVC K
	ARQ	DHS	
12.6015	CW	WLO	CQ CQ CQ DE WLO WLO AS
	ARQ	SEC	
12.602	ARQ	OXZ	
	ARQ	UAT	
12.60340	ARQ	KFS 1094	
12.6035	ARQ	GKY5	
	ARQ	LGB	TLX
12.604	ARQ	YUR	
	ARQ	WLO	(Receive on 12501.50)
12.6045	ARQ	WLO	(Receive on 12502.00)
12.605	ARQ	FFT2	
12.6055	ARQ	LGB	TLX
12.606	ARQ	WLO	(Receive on 12503.50)
12.6075	ARQ	WNU	DE WNU SELCAL 1109 12607R5/12505R5
12.608	ARQ	GKQ5	
12.6085	CW	WNU	DE WNU SELCAL 1109 12608R5/12505R5
12.61040	ARQ	WLO	
12.615	ARQ	KFS	1094
12.61530	CW	KPH	VVV DE KPH QSX 22 16 12 8 6 4 MHZ DE KPH 22 16 12 8 6 4 MHZ AR K

Frequency	Mode	Call Sign	Service / Times
12.640 MHz	CW	FUM	VVV DE FUM
	VOICE		SPY STATION, ENGLISH FEMALE VOICE, SENDS 5-DIGIT TRAFFIC, 0300, 0415 UTC
12.65785	CW	JNA	CQ CQ CQ DE JNA JNA JNA
12.65930	CW	WLO	CQ CQ CQ DE WLO WLO AS 3334 4257R5 6446R5 8445R5 8473R5 8658R0 12660R0 12704R5 13024R9 16968.5 17182R4 AND 22320R0 KHZ TFC LIST AND WX K
12.66240	CW	CBV	CQ DE CBV QSX CH 1/5/6 ON 4 6 8 12 MHZ AULTX 4 8 1216 SERIES 9 K
12.66390	CW	FUM	VVV DE FUM
12.67345	CW	JOU	CQ CQ CQ DE JOU JOU JOU QSX 12 MHZ K
12.67430	CW	CLA	CQ CQ CQ DE CLA CLA CLA
12.68175	CW	LGB	LGB TLX LGB TLX LGB TLX
12.68760	CW	OFJ	OFJ QSX 4 MHZ K OFJ QSX 8 MHZ K OFJ 12 MHZ K OFJ 16 MHZ K OFJ QSX 22 MHZ K
12.690	CW	PPJ	VVV DE PPJ PPJ PPJ QSX CH 4/5 K AND 22 MHZ QSW 22496 KHZ
	QSX		CH 2/3 K
	CW	UJY	CQ CQ DE UJY UJY QSS 8640/12690 ANS 8368.5/12552.5 K
12.6915	CW	FUX	VVV DE FUX
12.693	CW	URD	DE URD URD QSX 4/6/8/12
12.69480	CW	KFS	CQ DE KFS KFS KFS/B QSX 8 12 16 22 MHZ K
12.69530	CW	XSX	CQ CQ CQ DE XSX XSX XSX QSX 8/12/16/22 MHZ
12.6995	CW	HPP	VVV VVV VVV CQ CQ CQ DE HPP HPP HPP QTC ? AMVER ? OBS ? QSW RTG 500 KHZ / 4.275 / 8.589 / 12.699 MHZ CH 5/6/9 AND RTTY 12583.0 / 12.480.5 / 16.822.5 / 16.699.5 MHZ
12.700	CW	XSQ	CQ CQ CQ DE XSQ XSQ XSQ QRU ? (Republic of China)
12.702	CW	CKN	NAWS DE CKN II ZKR F1 2386 4168 6242.6 8341.5 12441.5 16650 22178 KHZ AR
12.70380	CW	WLO	DE WLO 3 OBS ? AMVERS ? QSX 6 8 12 16 25R071 MHZ NW ANS CH/6 K

Frequency	Mode	Call Sign	Service / Times
12.7085MHz	CW	CKN	NAWS DE CKN II ZKR F1 2386 4170 6251 8324 12386 16567 22191 KHZ K
12.709	CW	FJP	CQ CQ CQ DE FJP23 FJP23 FJP23 QRU QRU ? QSX 12 BAND K
12.7095	CW	8PO	DE 8PO
12.714	CW	UXN	DE UXN
12.7165	CW	NMN	CQ CQ CQ DE NMN NMN NMN QRU ? QSP AMVER/GOVT QSX H24 8/12/16 MHZ ITU CHANNS 4/5/6 QRU ? DE NMN NMN NMN K
12.71920	CW	ZLO	DE ZLO ZAY A1A 8 12 16 ZNI 1A 22 ZNI 1B J2B 8 12 J7B 12 AR
12.7215	CW	ZLO	DE ZLO ZAY A1A 6 8 12 ZNI 1A 8 12 ZNI 16 J2B 8 12 J7B 16 12 AR
12.727	CW	HLJ	CQ CQ CQ DE HLJ HLJ HLJ QSX 12 MHZ K
12.7275	CW	LGW	CQ CQ DE LGW LGU LGB LFN LGJ LGX LFX QSX 4G 6CG 8CQ 12CG 16CG AND 16740.7
12.728	FAX	USN	POINT REYES, CA, WEATHER
12.730	USB	NMC	U.S. COAST GUARD SAN FRANCISCO, CA
12.73655	CW	TAS	DE TAS TAS 12 MHZ CH 5 6 12 K
12.738	CW	PPR	VVV DE PPR PPR PPR QSX 12 MHZ K
12.73990	CW	ZLB	DE ZLB ZLB ZLB QSX 4 8 12 16 MHZ CH 5 6 17 22 MHZ CH 3 4 9
12.740	CW	HWN	VVV DE HWN
	CW	ZLB	DE ZLB2/4/5/6/7 ZLB2/3/4/5/6 QSX 4/8/12/16/22 MHZ CHLS 3/4/10 BT
12.741.85	FAX	GYA	NORTHWOOD, UNITED KINGDOM, 0000-2400 UTC
12.743	CW	NRV	CQ CQ CQ DE NRV NRV NRV QRU ? QSP AMVER QTC 3FUE2 12 MHZ CH 9/13 AND 16 MHZ CH 9/12 QRU ? DE NRV
12.745	VOICE		SPY STATION, ENGLISH FEMALE VOICE, SENDS 5-DIGIT TRAFFIC, 0130, 0430 UTC
12.74985	CW	CWA	VVV CQ CQ DE CWA CWA CWA QSX 4/6/8/12/16 MHZ

Frequency	Mode	Call Sign	Service / Times
12.750 MHz	CW	CKN	VVV VVV VVV DE CKN CKN CKN C13E C13E C13E
	USB		U.S. COAST GUARD, BOSTON, MA
	CW	NIK	VVV VVV VVV DE NIK NIK NIK
	FAX	NIK	BOSTON, MASS.
12.7505	CW	VSI	VVV VVV VVV DE VSI VSI VSI 4/5/6/7 AR
12.752	USB		U.S. COAST GUARD CUTTERS
	FAX		U.S. NAVY
12.754	USB	NMA	USCG, MIAMI, FLA
12.7625	CW	DAM	VVV DE DAM DAM
12.747	USB	MIW	ISRAELI MOSSAD INTELLIGNECE
12.7475	CW	CLS	CQ CQ CQ DE CLS CLS CLS QSW 12549/16732 KHZ QRU K
12.74925	CW	CWA	CQ CQ CQ DE CWA CWA CWA QSX 4/6/8/12/16 MHZ CH/6/16/18 AND 22 MHZ C3/4/6 K
12.751	FAX	NIK	BOSTON, MASS WEATHER
12.7524	CW	CKN	VVV VVV VVV DE CKN CKN CKN C13E C13E C13E
12.775	CW	XFU	CQ CQ CQ DE XFU XFU XFU
12.781	CW	D3E41	CQ DE D3E41/51/52/61 QSX 6 8 AND 12 MHZ C6 ON RTF CH 421 821 1621 BT AND 2207 AR K
12.7815	CW	9MB	VVV VVV VVV 9MB 6 13 16 19 (Malaysia)
	CW	YUR	VVV DE YUR 3/5 CH 3/4/7 K
12.78220	CW	OST	VVV DE OST4/42 OST5/52 ANS 8 OR 12 MHZ C
12.786	USB		U.S. COAST GUARD HONOLULU, HI
12.788	CW	JFA	CQ CQ CQ DE JFA JFA JFA K
12.79210	CW	CLA	CQ CQ CQ DE CLA CLA CLA QSX C/6 8364R4/12546R6/16728R8 TX 8573R0/12673R5/16976R1 QRJ CLA 20/32/41/50 K
12.79975	CW	YUR	VVV DE YUR QSX 12 MHZ CH 5/6/12 K
12.800	CW	PCH	DE PCH51 12 K
12.801	CW	TAS	DE TAS TAS 12 MHZ CH 5 6 12 K
12.8095	CW	KPH	VVV DE KPH QSX 22 16 12 8 6 4 MHZ DE KPH 22 16 12 8 6 4 K
12.822	CW	GKA	VVV VVV VVV DE GKA
12.8235	CW	CTP	DE CTP CTP QSX 4 6 8 12 MHZ AR

Frequency	Mode	Call Sign	Service / Times
12.8258MHz	CW	WNU34	CQ CQ CQ DE WNU34 WNU34 WNU34 QSX 4 6 8 12 16 MHZ OBS ?
12.827	CW	JCS	CQ CQ CQ DE JCS JCS JCS 12 MHZ K
12.8295	CW	XFM	CQ CQ CQ DE XFM XFM XFM QRU ? 4/8/12 VVV DE XFM XFM XFM QRU ? 4/8/12
12.8345	CW	GKB	DE GKB 1
12.838	CW	XDA	CQ CQ CQ DE XDA XDA XDA QRU ? K
12.8405	CW	JMC	CQ CQ CQ DE JMC/JMC2/JMC3/JMC5/JMC6
12.84380	CW	HLO	CQ CQ CQ DE HLO HLO HLO QSX 12 MHZ K
12.8445	CW	KFS	CQ DE KFS KFS KFS/A QSX 8 12 16 AND 22 MHZ K
	FAX	GZZ	LONDON, ENGLAND
12.84635	CW	WCC	VVV VVV DE WCC WCC BT OBS ? QSX 6 8 12 MHZ K
12.8495	CW	ZSJ	CQ CQ CQ DE ZSJ ZSJ ZSJ QSX AMVER 6/8/12/22 CH 3/4/9/10 AR K
12.850	RTTY	HKC	BUENAVENTURA, COLOUMBIA (Weather)
12.8533	CW	HKC	CQ CQ CQ DE HKC HKC HKC QRU ? QSX 8/12/16 MHZ ON CHANNELS 5/6 K
12.855	CW	UBF	4LT3 4LT3 DE UBF2 UBF2
12.85780	CW	6WW	VVV DE 6WW
12.860	CW	4XO	CQ DE 4XO QSX 8 C K
12.86370	CW	XSW	CQ CQ CQ DE XSW XSW XSW QRU ? QSX 8/12/16/22 MHZ K
12.869	CW	WNU54	CQ CQ CQ DE WNU54 WNU54 WNU54 QSX 4 6 8 12 16 MHZ OBS ?
12.87220	CW	XSG	CQ CQ CQ DE XSG XSG XSG QRU ? QSX 8 12 16 AND 22 MHZ BT
12.87340	CW	VCS	VVV VVV VVV CQ DE VCS VCS VCS QSX 4 8 AND 12 CHNL 11/5/6 K
12.875.25	CW	VAI	CQ CQ CQ DE VAI VAI VAI QSX 4/6/12/16 MHZ CH 4/5 OBS /AMVER/WESTREG/QRJ ? VAI SITOR SELCALL 1.00581 QRU ?
12.8785	CW	JCU	CQ CQ CQ DE JCU JCU JCU QSX 12 MHZ K
12.8795	CW	AAG	CQ DE AAG2/4/6 QSX 6 11
	CW	WSC	CQ CQ DE WSC WSC QSX 6 8 12 MHZ OBS ? DE WSC K

Frequency	Mode	Call Sign	Service / Times
12.8804 MHz	CW	SAG	CQ DE SAG2/4/6 QSX GROUP CH 7 AND COMMON CH 4 ON 4 8 12 MHZ BT FOR QRJ QSX 420 801 1203 AR
12.8865	FEC		WEATHER REPORTS
12.8875	CW	EAD	DE EAD2/EAD3/EAD44 QSX 8/12 MHZ CG AR K
12.8895	CW	NMO	CQ CQ CQ DE NMO NMO QRU AMVER IMI QSX 8/12 MHZ AMVER CHNL 5/6/11 ITU 22 MHZ AMVER CHNL 3/4 ITU AAA QLH 8650/12889.5 KHZ AND 22476 KHZ DE NMO NMO NMO QRU IMI K
	USB		U.S. COAST GUARD HONOLULU, HI
12.891	CW	UFN	DE UFN 22
12.899	CW	DAN	CQ CQ DE DAN DAN 8 CG 12 CG K
12.9005	CW	PWZ33	VVV VVV VVV DE PWZ33 PWZ33 PWZ33 ZZZ
12.903	FAX	MGR	ATHENS, GREECE, 0700-2000 UTC
12.907	CW	VHP	VVV VVV VVV DE VHP VHP VHP 2/3/4/5/6 AR
12.908	CW	VIX	VVV VVV VVV DE VIX VIX VIX
12.912	CW	FFL	CQ CQ CQ DE FFL2 FFL3 FFL4 FFL5 PSE QSX 6 MHZ 8 MHZ 12 MHZ K
12.9165	CW	KLB	CQ CQ CQ DE KLB KLB KLB QSX 4 6 8 12 18 AND 22 MHZ
	OBS		? SITOR SELCAL 1113 AR K
12.9255	CW	WCC	VVV VVV DE WCC WCC WCC BY OBS QSX 6 8 12 MHZ K
12.940	CW	LZJ	DE LZJ L25
12.94280	CW	RKLM	DE RKLM
12.958	CW	PPO	VVV DE PPO PPO PPO QSX CHANNELS 4/5 K
12.967	CW	A7D	VVV DE A7D
12.9685	CW	XSU	CQ CQ CQ DE XSU XSU XSU QRU QSX 8 12 AND 16 MHZ BK K
12.97030	CW	URL	CQ CQ DE URL URL ANS 12444.5 16666.5 PSE QSO K
12.975	CW	IQX	VVV DE IQX IQX IQX QSX 4 8 12 CG
12.97730	CW	URL	CQ CQ DE URL URL ANS 12446.5 16666.5 PSE QSO K
12.9785	CW	ICB	VVV DE ICB ICB ICB K 8 12 MHZ
	CW	FONIA	DE FONIA D/D QSA NIL QTC K

Frequency	Mode	Call Sign	Service / Times
12.9805 MHz	CW	PPL	VVV VVV VVV DE PPL PPL PPL QSX K
12.991	CW	WLO	DE WLO 3 OBS ? AMVERS ? QSX 8 12 16 25R071 MHZ NW ANS C9 K
12.994	CW	VIP04	VVV DE VIP04 VIP04 VIP04 QSX CH 5 6 ET 1 6
12.995	CW	UFL	DE UFL UFL UFL PSE 17110/8297.6 K

TIME CONVERSION CHART

U.T.C.	PST	PDST MST	MDST CST	CDST EST	EDST
0:00	4 pm	5 pm	6 pm	7 pm	8 pm
1:00	5 pm	6 pm	7 pm	8 pm	9 pm
2:00	6 pm	7 pm	8 pm	9 pm	10 pm
3:00	7 pm	8 pm	9 pm	10 pm	11 pm
4:00	8 pm	9 pm	10 pm	11 pm	Midnight
5:00	9 pm	10 pm	11 pm	Midnight	1 am
6:00	10 pm	11 pm	Midnight	1 am	2 am
7:00	11 pm	Midnight	1 am	2 am	3 am
8:00	Midnight	1 am	2 am	3 am	4 am
9:00	1 am	2 am	3 am	4 am	5 am
10:00	2 am	3 am	4 am	5 am	6 am
11:00	3 am	4 am	5 am	6 am	7 am
12:00	4 am	5 am	6 am	7 am	8 am
13:00	5 am	6 am	7 am	8 am	9 am
14:00	6 am	7 am	8 am	9 am	10 am
15:00	7 am	8 am	9 am	10 am	11 am
16:00	8 am	9 am	10 am	11 am	Noon
17:00	9 am	10 am	11 am	Noon	1 pm
18:00	10 am	11 am	Noon	1 pm	2 pm
19:00	11 am	Noon	1 pm	2 pm	3 pm
20:00	Noon	1 pm	2 pm	3 pm	4 pm
21:00	1 pm	2 pm	3 pm	4 pm	5 pm
22:00	2 pm	3 pm	4 pm	5 pm	6 pm
23:00	3 pm	4 pm	5 pm	6 pm	7 pm

(C) N6MQS

Frequency	Mode	Call Sign	Service / Times
13.001 MHz	CW	KPH	VVV DE KPH QSX 22 16 12 8 6 4 DE KPH QSX 22 16 12 8 6 4 K
13.003	USB		U.S. COAST GUARD PORTSMOUTH, MA
13.0039	FAX	CTU	WEATHER, MONSANTO
13.0085	CW	JOR	CQ CQ CQ DE JOR JOR JOR QSX 12 MHZ K
13.0115	CW	WNU44	CQ CQ CQ DE WNU44 WNU44 WNU44 QSX 4 6 8 12 MHZ OBS ?
13.015	CW	IAR	VVV VVV VVV DE IAR IAR IAR K 4 8 12 16 22
13.0193	CW	GKC	DE GKC
	CW	VPS60	CQ CQ DE VPS60 VPS60 VPS60
13.020	USB	NMF	USCG, BOSTON, MA
13.0225	CW	SPE	DE SPE QRL AS
13.024	CW	WLO	DE WLO 2 OBS ? AMVERS ? QSX 8 12 16 25R071 MHZ NW ANS C 5/6 QTC LIST FREQ 4343.0 6416.0 8140.0 12886.5 170225 22487.0 KHZ DE WLO 2 OBS ? AMVERS ? QSX 4 6 8 1216 22 25R172 MHZ NW ANS C 3/4 K
13.0304	CW	FUF	VVV DE FUF
13.033	CW	WCC	VVV VVV DE WCC WCC BT OBS ? QSX 6 8 12 MHZ K
13.037	CW	KLC	CQ CQ DE KLC KLC QSX 4 6 8 KHZ AND HF 4/6/8/12/16/22 MHZ CHS 5 6 9 KLC TFC LIST ON 1/2 HRS OBS ? AMVER ? QRU ? DE KLC K (Galveston, Texas)
13.0505	CW	UDK2	VLI4 VLI4 DE UDK2 UDK2 QSX 12476 K
13.051	CW	4XO	VVV DE 4XO QSX 12 C K
13.0543	CW	JDC	CQ CQ CQ DE JDC JDC JDC QSX 12 MHZ K
13.0555	CW	UJQ7	CQ CQ DE UJQ7 UJQ7 QSX 12560/16734 K
13.0562	CW	LSA	VVV VVV DE LSA 2/3/4/5 NW PRESS PSE AS BT
13.0624	CW	CLA	CQ CQ CQ DE CLA CLA CLA QSX C/6 8364R4/12546R6/16728R8 TX 8573/12673/16961 QSJ CLA 20/32/41/50 AR
13.064	CW	JDB	CQ CQ CQ DE JDB JDB JDB QSX 12 MHZ K
13.0664	CW	OST	VVV DE OST 4/42 OST 5/52 ANS 8 OR 12 MHZ C
13.0695	CW	JOS	CQ CQ CQ DE JOS JOS QSX 12 MHZ K
13.077	CW	JNA	CQ CQ CQ DE JNA JNA JNA
	USB		U.S. COAST GUARD, GUAM

Frequency	Mode	Call Sign	Service / Times
13.080 MHz	FEC		100 BAUD, AMTOR, RXREV ON WCC DE
		WCC	NOW ANS ATOR 4 6 8 12 16 22 MHZ QRU ? OBS AMVER DE WCC
13.082	USB		U.S. COAST GUARD HONOLULU, HI
13.083	USB	NMF	USCG, BOSTON, MA
	USB	KMI	
13.0845	USB	NMN	USCG, PORTSMOUTH, VA
13.0875	CW	HKB	CQ CQ CQ DE HKB HKB HKB QSX ON 8.374/12.546/16.728 MHZ K
13.089	USB		WEATHER FORECAST
13.0935	USB	WOM	AT&T SERVICES
13.0995	CW	WNU	DE WNU TFC SELCAL 11O914O9/13099R5/12519R5
13.105			VOICE PPR STANDARD TIME, BRAZIL
13.1132	USB	NMC	U.S. COAST GUARD
13.137.85	USB		VOICE WEATHER FOR NORTH ATLANTIC AREA
13.142	CW	CLA	CQ CQ CQ DE CLA CLA CLA QSX C/6 8364/12552/16736 TX 8573/12673R5/16961 QSW CLA 20/32/41/50
13.143.65	USB		SHIP TO SHORE VOICE
13.155	USB		SKYKING BROADCASTS
13.161	USB		MARINE SHIP TO SHORE
13.1615	CW	KMI	CQ CQ CQ DE KMI KMI KMI
13.1783	USB		NATIONAL WEATHER SERVICE
13.1865	CW	KMI	CQ CQ CQ DE KMI KMI KMI QRJ
	USB	KMI	SHIP TO SHORT PHONE PATCH
13.187.60	USB	KING5	FREQUENCY USED BY DRUG TRAFFICING
13.201	USB		USAF AIR/GROUND/SKYKING REPORTING
13.204	USB		AWACS EARLY WARNING
13.205	USB		SKYKING BROADCASTS
13.211	USB		SKYKING BROADCASTS
	USB		LOOKING GLASS
13.214	USB		USAF AIR/GROUND
	USB		MILITARY AIRLIFT COMMAND
	USB		USAF GLOBAL CONTROL AND COMMAND
13.217	USB		LOOKING GLASS
13.241	USB		USAF AIR/GROUND SAC
	USB		LOOKING GLASS
13.244	USB		USAF AIR/GROUND
	USB		USAF GLOBAL CONTROL AND COMMAND

Frequency	Mode	Call Sign	Service / Times
13.244 MHz	USB		SKYKING BROADCAST
13.247	USB		USAF TACTICAL AIR/GROUND
	USB		USAF GLOBAL CONTROL AND COMMAND
	USB		LOOKING GLASS
13.253	RTTY		BAGHDAD NEWS 150 SHIFT 50 BAUD, RXREV OFF
13.267	USB		NATIONAL HURRICANE SERVICE
13.270	USB		AERONAUTICAL WEATHER REPORTING
13.275	CW	CLP1	MINISTRY OF FOREIGN AFFAIRS, HAVANA, CUBA CW AND 50 BAUD RTTY
13.282	USB		WEATHER FORECAST
13.290	CW	CLP1	MINISTRY OF FOREIGN AFFAIRS, HAVANA, CUBA CW AND 50 BAUD RTTY
13.282	USB		AERONAUTICAL WEATHER REPORTING
13.288	USB		AERONAUTICAL WEATHER REPORTING
13.300	CW	CLP1	MINISTRY OF FOREIGN AFFAIRS, HAVANA, CUBA
13.306	USB		INTERNATIONAL AIRLINES
13.354	USB		NOAA HURRICANE HUNTERS
13.3615	ARQ	GPA 5	
13.3675	CW	KKN39	QRA QRA QRA DE KKN39 KKN39 KKN39 QSX 4/13/17/25 K (Will transmit call once every minute)
13.375	VOICE		SPY STATION, SPANISH FEMALE VOICE, SENDS 5-DIGIT TRAFFIC, 0700 UTC
13.377	RTY	CME338	(Cuba)
13.379	CW		SPY STATION, MORSE CODE, SENDS 5-DIGIT TRAFFIC, 0230, 0235, 1430 UTC
13.380	CQ	NRV	USCG COMMSTA
13.388.60	CW	CLP1	RPP RPP DE CLP1 QSV V'S (Ministry of Foreign Affairs,Cuba) (0150 UTC)
13.3875	CW	KKN39	QRA QRA QRA DE KKN39 KKN39 KKN39 QSX 4/13/17/25 K (Will transmit call once every minute)
13.4165	RTTY	CCS	50 BAUD, CHILE
13.4325	CW	KNY24	THAI, EMBASSY (Runs RTTY on this Frequency)

Frequency	Mode	Call Sign	Service / Times
13.435 MHz	VOICE		SPY STATION, SPANISH FEMALE VOICE, SENDS 5-DIGIT TRAFFIC, 0130, 0200, 0500 UTC
	CW		SPY STATION, MORSE CODE, SENDS 5-DIGIT TRAFFIC, 0000, 0030, 0040, 0045 UTC (WILL USE 14.375 ON HALF HOUR)
13.445	USB		SKY KING BROADCAST
13.450	VOICE		SPY STATION, SPANISH FEMALE VOICE, SENDS 4-DIGIT TRAFFIC, 0000, 0100, 2300 UTC
13.452	VOICE		SPY STATION, SPANISH FEMALE VOICE, SENDS 5-DIGIT TRAFFIC, 0000, 0030 UTC
13.454	FAX		HONG KONG, PRESS
13.479	USB		U.S. ARMY SPECIAL FORCES
13.470	FAX	RKU	MOSCOW, USSR, 0000-2400 UTC
13.486	USB		U.S. COAST GUARD CUTTERS
13.491	AM		VOICE OF AMERICA
13.504.85	USB		ENGLISH SPEAKING SPY TRANSMISSIONS
13.505	USB		U.S. ARMY MILITARY
13.506.5O	USB		U.S. ARMY MILITARY
13.510	USB	CIO	ISRAELI MOSSAD INTELLIGENCE
	FAX	CFH	HALIFAX, NOVA SCOTIA
13.520	ARQ		INTERPOL FREQUENCIES
13.526	CW		SPY STATION, MORSE CODE, SENDS 5-DIGIT TRAFFIC, 0530 UTC
13,545	VOICE		SPY STATION, GERMAN FEMALE VOICE, SENDS 5-DIGIT TRAFFIC, 0600, 0630 UTC
13.547	USB		SKYKING BROADCASTS
13.548	FAX	ZKLF	WELLINGTON, NEW ZEALAND, WEATHER, 0000-2400 UTC
13.550	CW	ZKLF	Sends 5-digit weather traffic
13.555	VOICE		SPY STATION, ENGLISH FEMALE VOICE, SENDS 3-2 DIGIT TRAFFIC, 1200 UTC
13.560	CW	BMB	VVV VVV VVV CQ CQ CQ DE BMB BMB BMB BT
13.595	FAX	JMH	TOKYO, JAPAN, WEATHER, 0000-2400 UTC
	VOICE		SPY STATION, SPANISH FEMALE VOICE, SENDS 5-DIGIT TRAFFIC, 0500 UTC
13.597	FAX	IMB	WEATHER, ROMA, ITALY

Frequency	Mode	Call Sign	Service / Times
13.600 MHz	USB		NASA MISSION FREQ
	LSB		U.S. MILITARY
13.605	AM		RADIO MOSCOW (English Service, East Coast) 2300-0400 UTC (May 5 - Aug 31) 0400-0530 UTC (May 5 - Aug 31) (West Coast) 0530-0800 UTC (May 5 - Sept 28) (West Coast)
	AM		RADIO AUSTRALIA, 0000-0100, 0900-1000, 1100-1200, 1600-2330 UTC
13.610	CW	C	HF BEACON LOCATED IN MOSCOW, USSR
	AM		RADIO DEUTSCHE WELLE, 0300-0600 UTC
13.625	AM		CHRISTIAN SCIENCE MONITOR, 1000-2255 UTC (Saturday), 1605-1755 UTC
13.630	AM		RADIO FOR PEACE INTERNATIONAL, 1800-0600 UTC
	AM		UNITED NATIONS RADIO, 2150-2200, 2345-0000 UTC, (Spanish Service), 1400-1415, 1530-1600 UTC
13.635	AM		SWISS RADIO INTERNATIONAL, (English Service), 0400-0430, 2100-2130 UTC, (Spanish Service), 2130-2200 UTC
	CW	K	"K" MARKER (Center Slot on Swiss International)
	CW	S	"S" MARKER
	CW	F	"F" MARKER
	AM		RED CROSS BROADCASTING SERVICE, (Mondays), 1040-1057 UTC, 1310-1327, 0310-0327 UTC
13.645	ARQ		DEPARTMENT OF STATE IN WASHINGTON, D.C.
	AM		RADIO MOSCOW, (English Service, West Coast), 0400-0530 UTC, (May 5 thru Aug 31), 0530-0800 UTC, (May 5 thru Sept 28)

Frequency	Mode	Call Sign	Service / Times
13.645 MHz	AM		RADIO MOSCOW WORLD SERVICE, (English Service) 1500-2200 UTC
13.653	AM		RADIO EGYPT
13.650	AM		RADIO AUSTRALIA, 1400 UTC
13.655	AM		THE VOICE OF JORDAN, (English Service), 1200-1415 UTC
	AM		RADIO FOR FREE IRAQ, 0400 UTC (Heavy Jamming)
	AM		BRUSSELS CALLING, (English Service), 0030-0130, (Spanish Service), 0000-0030 UTC
13.660	AM		THE VOICE OF BAGHDAD, (English Service) 2100-2300 UTC
	AM		BRITISH BROADCASTING COKPR., 2130-2200 UTC
13.665	AM		RADIO MOSCOW, (West Coast), 0630-0900 UTC, (Mar thru Nov)
13.670	AM		CBC RADIO, CANADA, 2200 UTC
	AM		RADIO SOUTH KOREA, (KBS), (English Service), 0800-0900, (Spanish Service), 1015-1100 UTC
	AM		RADIO CANADA INTERNATIONAL (Mon. thru Fri.) 2200-2230, 2330-0030 UTC
13.673	CW	KRH51	U.S. DEPT OF STATE, LONDON, GREAT BRITAIN CW AND 50, 75 BAUD RTTY
13.675	AM		THE VOICE OF THE UNITED EMIRATES, (English Service)
13.685	AM		SWISS RADIO INTERNATIONAL, 1530-1600 UTC
13.6895	ARQ		ISRAEL EMBASSY

Frequency	Mode	Call Sign	Service / Times
13.695 MHz	AM		RADIO FRANCE, 0300-0700 UTC
	AM	WYFR	FAMILY RADIO NETWORK, 1300-1400 UTC, (English to Europe), 0500-0800 UTC
13.700	AM		RADIO HAVANA, CUBA, (English Service), 2000-2200 UTC, (Spanish Service), All other Broadcast Times.
13.705	AM		RADIO AUSTRALIA, 2100-2300 UTC
	AM		RADIO MOSCOW, 0400 UTC
13.710	AM		BRUSSELS CALLING, (English Service), 0030-0130, (Spanish Service), 0000-0030 UTC
13.720	AM		RADIO CHINA, (English Service) 0800-1000 UTC
13.730	AM		RADIO AUSTRIA INTERNATIONAL, 0130-0200, 0330-0400, 0530-0600, 0830-0900, 1130-1200, 1530-1600, 1830-1900 UTC, (English Broadcasts World Wide), (Spanish Service), 0230-0330 UTC
13.740	AM		VOICE OF AMERICA, (Spanish to American Republics) 0100-0400, 0930-1130 UTC
13.742	USB		NASA MISSION FREQ
13.747	ARQ		INTERPOL FREQUENCIES
13.755	AM		RADIO AUSTRALIA, 1430-2100 UTC
13.760	AM		VOICE OF AMERICA, 0300-0500 UTC
	AM		WHRI RADIO INTERNATIONAL, 1900-2330 UTC

Frequency	Mode	Call Sign	Service / Times
13.760 MHz	AM		CHRISTIAN SCIENCE MONITOR, 0000-0455, 0800-0855, 1200-1455 (Occasional Use Times), 1000-1200 UTC, (Spanish Service), (Sabado, Domingo), 0405-0455, 1405-1455 UTC
13.770	AM		VOICE OF AMERICA, 2000-2200 UTC
	AM		RADIO DEUTSCHE WELLE, 0300-0500 UTC
	AM		CHRISTIAN SCEINCE MONITOR, 2000-2055 UTC
13.775	AM		VOICE OF AMERICA, (Spanish to American Republics) 1200-1500 UTC
13.781	RTTY		NORTH KOREAN NEWS AGENCY
13.783	CW	KRH51	U.S. DEPT OF STATE, LONDON, GREAT BRITAIN CW AND 50, 75 BAUD RTTY
13.785	AM		RADIO NEW ZEALAND, 1900-2400 UTC
	AM		RED CROSS BROADCASTING SERVICE, (English), 0740-0757 UTC
13.8025	RTTY	RCR	RUSSIAN WEATHER TRAFFIC, 5 DIGIT GROUPS
13.815	CW	KRH50	QRA QRA QRA DE KRH50 KRH50 KRH50 QSX 5/7/11/13/16/20 ? QSX 5/7/11/13/16/20 K (Will transmit call once every minute)
13.820	ARQ		INTERPOL FREQUENCIES
13.826	USB		US NAVY MARS
13.850	RTTY/CW		VIET NAM EMBASSY TO YEMEN
13.855	AM		VOICE OF AMERICA, (Relay Link Times Vary)
	AM		ICELANDIS NATIONAL BROADCASTING SERVICE, (Icelandic Service) 1410-1440, 1935-2010, 2300-2335 UTC
13.860	AM		VOICE OF AMERICA, (Relay Link Times Vary)
13.878	USB		NASA MISSION FREQ

Frequency	Mode	Call Sign	Service / Times
13.885 MHz	AM		VOICE OF AMERICA, (Relay Link Times Vary)
13.907	USB		SKYKING BROADCASTS
13.918	FAX	AXM35	CANBERRA, AUSTRALIA, WEATHER, 0000-2400 UTC
13.922	VOICE		SPY STATION, ENGLISH FEMALE VOICE, SENDS 5-DIGIT TRAFFIC, 2200 UTC
13.931	CW		SPY STATION, MORSE CODE, SENDS 2 GROUPS OF 4-DIGIT TRAFFIC, 0330 UTC
13.9755	USB	NPG	NAVAL COMMUNICATIONS STATION, STOCKTON, CA
13.977	CW	CLP7	CUBAN
13.986	CW		U.S. AIR FORCE MARS
13.995	AM		VOICE OF AMERICA
13.9975	USB		AIR HF/MARS 2045TH COMMUNICATIONS GROUP, ANDREWS AFB, WASHINGTON, DC

TIME CONVERSION CHART

U.T.C.	PST	PDST MST	MDST CST	CDST EST	EDST
0:00	4 pm	5 pm	6 pm	7 pm	8 pm
1:00	5 pm	6 pm	7 pm	8 pm	9 pm
2:00	6 pm	7 pm	8 pm	9 pm	10 pm
3:00	7 pm	8 pm	9 pm	10 pm	11 pm
4:00	8 pm	9 pm	10 pm	11 pm	Midnight
5:00	9 pm	10 pm	11 pm	Midnight	1 am
6:00	10 pm	11 pm	Midnight	1 am	2 am
7:00	11 pm	Midnight	1 am	2 am	3 am
8:00	Midnight	1 am	2 am	3 am	4 am
9:00	1 am	2 am	3 am	4 am	5 am
10:00	2 am	3 am	4 am	5 am	6 am
11:00	3 am	4 am	5 am	6 am	7 am
12:00	4 am	5 am	6 am	7 am	8 am
13:00	5 am	6 am	7 am	8 am	9 am
14:00	6 am	7 am	8 am	9 am	10 am
15:00	7 am	8 am	9 am	10 am	11 am
16:00	8 am	9 am	10 am	11 am	Noon
17:00	9 am	10 am	11 am	Noon	1 pm
18:00	10 am	11 am	Noon	1 pm	2 pm
19:00	11 am	Noon	1 pm	2 pm	3 pm
20:00	Noon	1 pm	2 pm	3 pm	4 pm
21:00	1 pm	2 pm	3 pm	4 pm	5 pm
22:00	2 pm	3 pm	4 pm	5 pm	6 pm
23:00	3 pm	4 pm	5 pm	6 pm	7 pm

Frequency	Mode	Call Sign	Service / Times
14.000 MHz	**SSB/CW**		**START OF AMATEUR RADIO 20 METER BAND (Ends 14350.00)**
14.0475	CW	W1AW	CW BULLETINS
14.075	RTTY		TELETYPE AMATEUR FREQUENCY
14.095	RTTY		ARRL RTTY BULLETINS
	PACKET	HH2OK	PACKET BULLETIN BOARD SYSTEM, HAITI WEST, INDIES
14.100	PACKET		AMATEUR RADIO PACKET NETS
	CW		20 METER BEACON, KH6O
14.170	VOICE		SPY STATION, ENGLISH FEMALE VOICE, SENDS 5-DIGIT TRAFFIC, 0400, 0405, 0410 UTC
	RTTY/CW		SPY STATION, MORSE AND RTTY, SENDS 5-DIGIT 0300, 0400, 0500 UTC
14.187	ARQ		INTERPOL FREQUENCIES
14.230	SSTV		INTERNATIONAL SLOW SCAN TELEVISION FREQUENCY
14.245	USB		AMATUER FAX
14.260	VOICE		SPY STATION, SPANISH FEMALE VOICE, SENDS 5-DIGIT TRAFFIC, 0400 UTC
14.275	USB		AMATEUR RADIO EMERGENCY INTERNATIONAL NET
14.280	CW		SPY STATION, MORSE CODE, SENDS 5-DIGIT TRAFFIC, 0600 UTC
14.282	USB		AMATEUR AMSAT NET
14.290	USB	W1AW	ARRL VOICE BULLETIN
14.320	CW		SPY STATION, MORSE CODE, SENDS 5-DIGIT TRAFFIC, 0600 UTC
14.347	USB		QUARTER CENTURY WIRELESS ASSOCIATION, 2000 UTC, SUNDAY
14.360	CW	KWS78	QRA QRA QRA DE KWS78 KWS78 KWS78 QSX 3/4/7/10/14 K (Will transmit call once every minute) (Department of State, Athens, Greece)
	VOICE		SPY STATION, SPANISH FEMALE VOICE, SENDS 5-DIGIT TRAFFIC, 0100, 0330 UTC
14.363	FAX	BAF	BEIJING, PEOPLES REPUBLIC OF CHINA, WEATHER
14.365	AM		VOICE OF AMERICA

Frequency	Mode	Call Sign	Service / Times
14.3708MHz	TDM		99 BAUD, TDM 2:4 RXREV OFF
14.3725	USB		U.S. AIR FOCRE
14.373	RTY	YIL71	IRAQ NEWS AGENCY
	USB	NPG	NAVY, SAN FRANCISCO, CA
14.375	VOICE		SPY STATION, SPANISH FEMALE VOICE, SENDS 5-DIGIT TRAFFIC, 0100, 0500 UTC
	CW		SPY STATION, MORSE CODE, SENDS 5-DIGIT TRAFFIC, 0000 UTC (WILL REPEAT AT 0030 UTC ON 13.435)
14.380	LSB		CUBIAN MILITARY
14.383	USB		U.S. NAVY MARS
14.384	USB		U.S. NAVY MARS
14.3885	USB		PLO NET OF EL FATAH
14.390	VOICE		SPY STATION, ENGLISH FEMALE VOICE, SENDS 3-2 DIGIT TRAFFIC, 1500 UTC
14.392	USB		AIR USAF WASHINGTON, DC
14.398	AM		VOICE OF AMERICA
14.3895	USB	NAV	HQ NAVY-MARINE CORPS MARS RADIO STATION, CHELTENHAM, MD
14.3905	USB	AFA	USAF, ANDREWS AFB, MD
14.3915	USB		U.S. NAVY MARS
14.400	USB	NAM	NAVAL COMMUNICATIONS AREA MASTER STATION, NORFOLK, VA
14.402	USB	WAR	HF/MARS/ARMY, FORT RITCHIE, MD
14.4025	CW		FORT HUACHUCA, ARIZONA MARS NET, 0300 UTC
14.4035	USB	WAR	HF/MARS RADIO STATION, FT DETRICK, MD
14.405	VOICE		SPY STATION, ENGLISH FEMALE VOICE, SENDS 3-2 DIGIT TRAFFIC, 1300 UTC
14.4065	USB		U.S. ARMY MARS
14.408	USB		U.S AIR FORCE MARS
14.410	CW	Y7F47	IRAQ, EMBASSY
14.411	USB		U.S. AIR FORCE MARS
	VOICE		SPY STATION, SPANISH FEMALE VOICE, SENDS 5-DIGIT TRAFFIC, 0300, 0330 UTC
14.415	VOICE		SPY STATION, SPANISH FEMALE VOICE, SENDS 5-DIGIT TRAFFIC, 0400, 0700, 2300 UTC

Frequency	Mode	Call Sign	Service / Times
14.417 MHz	RTTY		IRAN NEWS AGENCY
14.422	RTTY		KUWAIT NEWS SERVICE
14.423	VOICE		SPY STATION, ENGLISH FEMALE VOICE, SENDS 3-2 DIGIT TRAFFIC, 1400 UTC
14.430	VOICE		SPY STATION, SPANISH FEMALE VOICE, SENDS 5-DIGIT TRAFFIC, 0530 UTC
	CW		SPY STATION, MORSE CODE, SENDS 5-DIGIT TRAFFIC, 0600 UTC
14.436	FAX	GFE	BRACKNELL, UNITED KINGDOM, 0000-2400 UTC
14.440	USB		U.S. ARMY MARS
	VOICE		SPY STATION, ENGLISH FEMALE VOICE, SENDS 5-DIGIT TRAFFIC, 1700 UTC SENDS 4-DIGIT TRAFFIC, 1300 UTC
14.441	USB		U.S. NAVY MARS
14.443	USB		U.S. NAVY MARS
14.445	USB		U.S. NAVY MARS
14.448	USB	AFC24	USAF, WESTOVER, MA
14.450	USB		PORTUGAL COMMUNICATIONS LINK
14.460	CW		SPY STATION, MORSE CODE, SENDS 5-DIGIT TRAFFIC, 0500 UTC
14.4635	USB		U.S. NAVY MARS
14.464.02	RTTY		50 BAUD, BAUDOT, RXREV OFF
14.4665	USB		U.S. NAVY MARS
14.467	USB		U.S. ARMY MARS
14.4688	USB		U.S. NAVY MARS
14.470	USB		U.S. NAVY MATS
14.476	USB		U.S. AIR FORCE MARS
14.477	USB		U.S. NAVY MARS
14.4783	TDM		97 BAUD TDM 2:4
14.480	RTTY		TEL AVIV, ISRAEL
14.483.07	AMTOR		99 BAUD, AMTOR, RXREV OFF
14.4835	USB		U.S. NAVY MARS
14.485	USB		U.S. NAVY MARS
14.4855	USB		U.S. ARMY MARS
14.487	USB		U.S. ARMY MARS
	VOICE		SPY STATION, ENGLISH FEMALE VOICE, SENDS 5-DIGIT TRAFFIC, 1200, 1300, 1400 UTC
14.4875	RTTY		RADIO HAVANA CUBA, 50 BAUD BAUDOT

Frequency	Mode	Call Sign	Service / Times
14.4885MHz	USB	AAH	HF/MARS RADIO STATION, FT LEWIS, WA VVV VVV VVV DE AAH FT LEWIS WA
14.505	USB		U.S. AIR FORCE
14.5115	USB	AAE	HF/MARS RADIO STATION, FT SAM HOUSTON, TX
14.525	CW		SPY STATION, MORSE CODE, SENDS 5-DIGIT TRAFFIC, 0800, 1000 UTC
14.526	AM		VOICE OF AMERICA
14.527	USB		HF/MARS USAF, HICKMAN, HI
14.5295	USB		U.S. AIR FORCE MARS
14.530	USB		U.S. AIR FORCE MARS
14.540	VOICE		SPY STATION, SPANISH FEMALE VOICE, SENDS 4-DIGIT TRAFFIC, 0430 UTC
14.545	VOICE		SPY STATION, SPANISH FEMALE VOICE, SENDS 5-DIGIT TRAFFIC, 0400 UTC
14.5475	RTTY		NAU NACCOMMSTA, SAN JUAN, PR
14.550	CW		DE SPW QSX 12525.5 KHZ
14.560	RTTY		JORDAN NEWS AGENCY
	VOICE		SPY STATION, SPANISH FEMALE VOICE, SENDS 5-DIGIT TRAFFIC, 0530 UTC
14.570	RTTY		NORTH KOREAN NEWS AGENCY
14.582	FAX	GFA	BRACKNELL, UNITED KINGDOM, 0600-1800 UTC
14.585	USB		NASA MISSION FREQ
14.5925	LSB		CUBAN MILITARY
14.605	ALIST		US AIF FORCE, (5-digit coded traffic)
	CW		SPY STATION, MORSE CODE, SENDS 5-DIGIT TRAFFIC, 0400 UTC
14.606	USB		U.S. AIR FORCE MARS
14.607	ARQ		INTERPOL FREQUENCIES
14.608	FAX	RCH	RUSSIAN METEOROLOGICAL SERVICE
14.616	CW		SPY STATION, MORSE CODE, SENDS 5-DIGIT TRAFFIC, 0405 UTC
14.623	ARQ		INTERPOL FREQUENCIES
14.625.82	TDM		99 BAUD, TDM 1:8 RXREV OFF
14.630	RTTY		50 BAUD, BAUDOT, RXREV OFF
	VOICE		SPY STATION, RUSSIAN MALE VOICE, SENDS 5-DIGIT TRAFFIC, 1700 UTC
14.638	CW		SPY STATION, MORSE CODE, SENDS 5-DIGIT TRAFFIC

Frequency	Mode	Call Sign	Service / Times
14.652 MHz	VOICE		SPY STATION, ENGLISH FEMALE VOICE, SENDS 5-DIGIT TRAFFIC, 1700 UTC
14.654	CW	IPW	EEE's (Long Series of E's) DE IPW QSX 12.5765 KHZ
14.655	CW	KRH51	U.S. DEPT OF STATE, LONDON, GREAT BRITAIN. CW AND 50, 75 BAUD RTTY
14.665	USB		U.S. ARMY MARS
14.686	USB		DRUG ENFORCEMENT AGENCY
14.670	VOICE	CHU	OTTAWA CANADA, STANDARD TIME/FREQUECNY
14.690	FAX	JMJ	JAPAN WEATHER
14.691	RTTY	CME	HAVANA CUBA, 6-BIT RTTY
14.694	CW		SPY STATION, MORSE CODE, SENDS 5-DIGIT TRAFFIC, 0400 UTC
14.699	RTY	YIX70	IRAQ NEWS AGENCY
14.703	RTTY	PCW	RTTY FROM DESERT STORM, BAGHDAD, IRAQ
	VOICE		SPY STATION, ENGLISH FEMALE VOICE, SENDS 3-2 DIGIT TRAFFIC, 1400, 1500, 1530, 1600 UTC
14.704	FAX	AOJ	SPANISH WEATHER
14.707	ARQ		INTERPOL FREQUENCIES
14.716	USB		SKYKING BROADCASTS
14.717	RTTY	JAE	KYODO PRESS, TOKYO NEWS AGENCY, (JAE58/JAT28)
14.730	CW	CLP1	MINISTRY OF FOREIGN AFFAIRS, HAVANA, CUBA CW AND 50 BAUD RTTY
14.735	RTTY	BZT	CHINA NEWS/WEATHER (SHIFT 450, 50 BAUDOT)
14.737	FAX	RXO	RUSSIAN METEOROLOGICAL SERVICE
14.740	VOICE		SPY STATION, ENGLISH FEMALE VOICE, SENDS 3-2 DIGIT TRAFFIC, 1300, 1400 UTC
14.744	USB		SKYKING BROADCASTS
14.745	USB		SKYKING BROADCASTS
14.7485	ARQ		SAUDI ARABIA EMBASSY IN CANADA
14.759.85	USB		U.S. AIR FORCE MARS
14.7605	ARQ		U.S. NAVY MARS

Frequency	Mode	Call Sign	Service / Times
14.761 MHz	CW	X1W	MEXICO
	CW	E9T	SENDS CODE TRAFFIC PLUS RTTY CODED TRAFFIC, TRANSMITS ON 08119 AT SAME TIME
14.762	CW		SPY STATION, MORSE CODE, SENDS 5-DIGIT TRAFFIC, 0400 UTC
14.770	VOICE		SPY STATION, SPANISH FEMALE VOICE, SENDS 5-DIGIT TRAFFIC, 0500 UTC
14.7745	USB		SKYKING BROADCASTS
14.787	CW		SPY STATION, MORSE CODE, ARABIC, 5-DIGIT SPECIAL LETTERS, 0400 UTC
14.800	AM		VOICE OF AMERICA
14.8011	TDM		99 BAUD, TDM 1:4, RXREV OFF
14.8135	USB		U.S. MILITARY
14.816	VOICE		SPY STATION, SPANISH FEMALE VOICE, SENDS 5-DIGIT TRAFFIC, 0400 UTC
14.817	ARQ		INTERPOL FREQUENCIES
14.8175	LSB		100 BAUD ARQ, INTERPOL
14.818	CW	CLP1	MINSTRY OF FOREIGN AFFAIRS, HAVANA, CUBA, CW AND 50 BAUD RTTY
	ARQ		INTERPOL FREQUENCIES
14.822	CW	CLP7	CUBAN
14.828	FAX	NPM	PEARL HARBOR, HI, WEATHER
14.829	USB		U.S. AIR FORCE MARS
14.835	VOICE		SPY STATION, SPANISH FEMALE VOICE, SENDS 5-DIGIT TRAFFIC, 0630, 0700 UTC
14.840	USB		U.S. NAVY MARS
14.842	FAX	ATP	NEW DELHI, INDIA, 0230-1400 UTC
14.846	VOICE		SPY STATION, ENGLISH FEMALE VOICE, SENDS 5-DIGIT TRAFFIC, 0500 UTC
14.858	VOICE		SPY STATION, RUSSIAN MALE VOICE, SENDS 5-DIGIT TRAFFIC, 1800 UTC
14.862	USB		U.S. AIR FORCE
14.864	VOICE		SPY STATION, SPANISH FEMALE VOICE, SENDS 5-DIGIT TRAFFIC, 0400 UTC
14.877	USB		U.S. ARMY MARS
14.878	USB		U.S. AIR FORCE MARS
14.8785	RTTY	JMG	JAPAN WEATHER TRAFFIC
14.880	AQR		DEPARTMENT OF STATE IN WASHINGTON, D.C.

Frequency	Mode	Call Sign	Service / Times
14.8865 MHz	TDM		U.S. NAVY TDM 2:4
14.894	USB		USAF NORAD HQ.
14.896	USB		NASA MISSION FREQ
14.905	USB		CIVIL AIR PATROL
14.926	RTTY		80 BAUD, 6-BIT RTTY, RXREV OFF
14.9268	RTTY	BUA	WEATHER REPORTS FROM PEOPLE'S REPUBLIC OF CHINA
14.930	VOICE		SPY STATION, GERMAN FEMALE VOICE, SENDS 5-DIGIT TRAFFIC, 1200 UTC (WILL ALSO USE 11.190)
14.934	USB		U.S. NAVY MARS
14.938	USB		U.S. ARMY MARS
14.9545	USB		MILITARY MARS
14.955	USB		SKYKING BROADCASTS
14.9575	AMTOR		100 BAUD, AMTOR, RXREV OFF
14.9665	CW		CUBAN MILITARY CW NET, 0800 UTC
14.970	VOICE		SPY STATION, ENGLISH FEMALE VOICE, SENDS 3-2 DIGIT TRAFFIC, 0400 UTC
14.982	FAX	RBV	RUSSIAN WEATHER, (When not sending traffic, will send continuous series of dit's)
14.985	AM		VOICE OF AMERICA
14.995	CW	RWU	VVV CQ CQ CQ DE RWU RWU (Timing Signals from the USSR)

DATE	FREQ.	MODE	TIME	STATION	SIGNAL	COMMENTS	SWL SENT	SWL REC'D

Frequency	Mode	Call Sign	Service / Times
15.000 MHz	**VOICE**	**WWV**	**INTERNATIONAL STANDARDS TIME AND FREQUENCY**
	VOICE		WWVH
	VOICE	BSF	REPUBLIC OF CHINA
	VOICE	JJY	TIMING SIGNAL FROM JAPAN
	CW	RTA	TIMING SIGNAL FROM U.S.S.R.
15.004	CW	RID	TIMING SIGNAL FROM U.S.S.R.
15.010	AM		RADIO "THE VOICE OF VIETNAM", (English Service), 1000-1030, 1230-1300, 1330-1400, 1600-1630, 1800-1830, 1900-1930, 2030-2100, 2330-2400 UTC, (Spanish Service), 1100-1130, 2000-2030 UTC
15.014	USB		USAF MIDDLE EAST
15.015	USB		USAF AIR/GROUND
	USB		MILITARY AIRLIFT COMMAND
	USB		USAF GLOBAL CONTROL AND COMMAND
15.024	CW		RUSSIAN AIRLINE AEROFLOT
15.030	AM		RADIO FOR PEACE INTERNATIONAL, 0000-0900, 1200-1800 UTC
	AM		UNITED NATIONS RADIO, 2150-2200, 2345-0000 UTC, (Spanish Service), 1400-1415, 1530-1600 UTC
15.035	USB		CANADIAN MILITARY WEATHER
	USB		SKYKING BROADCASTS
15.041	USB		USAF SAC AIR/GROUND
	USB		SKYKING BROADCASTS
15.044	USB		LOOKING GLASS
15.060	AM		RADIO SAUDI ARABIA, 0400 UTC
	VOICE		SPY STATION, ENGLISH FEMALE VOICE, SENDS 5-DIGIT TRAFFIC, 1400 UTC
15.070	AM		BRITISH BROADCASTING CORPORATION, 0300-2330 UTC
15.091	USB		SKYKING BROADCASTS
	USB		AWACS EARLY WARNING

Frequency	Mode	Call Sign	Service / Times
15.105 MHz	AM		VOICE OF AMERICA, (Service to South Asia), 0130-0300 UTC, (Middle East), 0300-0430 UTC
	AM		BRITISH BROADCASTING CORPORATION, 0800-0830, 1300-1345 UTC
	AM		RADIO YUGOSLAVIA, 2100 UTC
15.110	AM		RADIO EXTERIOR DE ESPANA, (Spanish Service) 1900-2300 UTC
15.115	AM		VOICE OF AMERICA, ALSO ON 06035.00, 09575.00, (Service to Africa), 0300-0700 UTC, (Europe), 1800-1830 UTC
	AM		RADIO FREE EUROPE (Bulgarian), 1300-1800 UTC
	AM	HCJB	QUITO, ECUADOR 0600-0700 UTC
	AM	HCJB	THE VOICE OF THE ANDES (English Service) 1130-1430 UTC
15.120	AM		VOICE OF AMERICA, (Service to Caribbean), 1000-1200, 2130-2200 UTC, (Haiti), 2100-2130, 2200-2300 UTC, (Latin America) 2130-2200 UTC
	AM		RADIO NEW ZEALAND INTERNATIONAL, 1800-2206 UTC (Mon to Sat)
15.125	AM		VOICE OF AMERICA, (Service to South Asia), 1330-1700 UTC, (Indo-China), 1100-1230 UTC, (South East Asia), 1230-1300 UTC

15

Frequency	Mode	Call Sign	Service / Times
15.130 MHz	AM		RADIO FREE EUROPE (Latvian), 1500-1600 UTC, (Lithuanian) 1600-1700 UTC, (Estonian), 1800-1900 UTC, (Latvian), 1900-2000 UTC, (Lithuanian), 2000-2100 UTC, (Estonian), 2100-2200 UTC
	AM		RADIO BEIJING CHINA, 0000 UTC
	AM	WYFR	FAMILY RADIO NETWORK, (Spanish Service), 1100-1500 UTC
15.140	AM		RADIO CUBA, 0000-0200 UTC (English service)
	AM		RADIO MOSCOW, 0200-0400 UTC
	AM		RADIO CHILE, 2000-2200 UTC
	AM		BRITISH BROADCASTING COPR., 2130-2200 UTC
	AM	HCJB	THE VOICE OF THE ANDES (Spanish Service) 1500-0500 UTC
15.145	AM		RADIO FREE EUROPE (Tatar-Bashkir) 0300-0600 UTC
	AM		WINB, RED LION, PA, 0100 UTC
	AM		VOICE OF AMERICA, (Service to Middle East), 1800-2100 UTC
	AM	WYFR	FAMILY RADIO NETWORK, (Spanish Service), 1100-1200 UTC

Frequency	Mode	Call Sign	Service / Times
15.150 MHz	AM		RADIO AUSTRALIA, 1400 UTC
	AM		VOICE OF AMERICA, (Swahili to Africa), 1630-1730 UTC
15.155	AM		VOICE OF AMERICA (English to Pacific service) 1100-1400 UTC
	AM	HCJB	QUITO, ECAUADOR, 0000-0600 UTC
	AM	HCJB	THE VOICE OF THE ANDES (English Service) 0030-0430 UTC
15.160	AM		VOICE OF AMERICA (English to Pacific service) 1400-1500 UTC (English to MIddle East/Europe service) 0800-1200 UTC (English to South Asia) 0100-0300 UTC (English to VOA Europe) 0300-0330, 0800-1000 UTC
	AM		RADIO AUSTRALIA, 2100-0700, 0800-1100 UTC
	AM		ORGANIZATION OF AMERICAN STATES, All programing in English except for Spanish, 2345-0030 UTC
	AM		UNITED NATIONS RADIO, (Spanish Service), 0015-0020 UTC
15.165	AM		VOICE OF AMERICA, (Service to Africa), 1800-1900 UTC
	AM		RADIO NORWAY INTERNATIONAL, 2000 UTC

Frequency	Mode	Call Sign	Service / Times
15.170 MHz	AM	WYFR	FAMILY RADIO NETWORK, (Spanish Service), 2300-0200 UTC
	AM		RADIO FREE EUROPE, (Azerbaijan) 0400-0500, 1100-1200 UTC, (Armenian) 1200-1400 UTC, (Azerbaijan) 1400-1500 UTC, (Georgian) 1500-1700 UTC, (Azerbaijan) 1700-1800 UTC, (Georgian) 1800-1900 UTC, (Armenijan) 1900-2000 UTC
	AM		RADIO AUSTRALIA, 0500-0600, 0900-1000, 1100-1200
	AM		RADIO SOUTH KOREA, (KBS), 0600-0700 UTC
15.175	AM		RADIO MOSCOW WORLD SERVICE
15.180	AM		RADIO MOSCOW, (English Service, West Coast), 0400-0800 UTC, (May 5 thru Sept 28)
	AM		VOICE OF AMERICA, (English to Pacific service) 1900-2000 UTC
15.185	AM		VOICE OF AMERICA, (English to Pacific service) 2100-0100 UTC (Spanish to American Republics) 1700-1730 UTC
15.190	AM		BRITISH BROADCASTING CORPORATION, WORLD SERVICE, (English Service), 0900-1100 UTC, (Spanish Service), 1100-1130 UTC
	AM		RADIO BANGLADESH, 1200 UTC

Frequency	Mode	Call Sign	Service / Times
15.195 MHz	AM		VOICE OF AMERICA, (English to Middle East/Europe service) 0800-1000 UTC (English to VOA Europe), 0800-1000 UTC
	AM		RADIO JAPAN 0100-0200, 0300-0400, 0500-0600, 0700-0800, 2300-0000 UTC
15.205	AM		VOICE OF AMERICA, (English to American Republics service) 0000-0230 UTC (English to Middle East/Europe service) 0500-0700,1400-2100 UTC (Russian to USSR) 1200-1400 UTC
	AM		RADIO DEUTSCHE WELLE, 0300-0350 UTC
	AM		BRITISH BROADCASTING CORP., 1100-1630 UTC (Canada & U.S.)
	ARQ		INTERPOL FREQUENCIES
15.215	AM		RADIO FREE EUROPE, (Romanian), 0700-1800 UTC
	AM		VOICE OF AMERICA, (Service to South East Asia), 0000-0030, (USSR), 1800-2100 UTC, (Middle East), 2200-2400 UTC
	AM	WYFR	FAMILY RADIO NETWORK, 1500-1700 UTC, (Spanish Service) 2300-0100 UTC
15.220	AM		RADIO NORWAY INTERNATIONAL, 1900 UTC
	AM		BRITISH BROADCASTING COPR., 1430-1630 UTC

Frequency	Mode	Call Sign	Service / Times
15.225 MHz	AM		VOICE OF AMERICA, (English to Middle East/Europe service), 0000-0300, 1000-1400, 2200-2400 UTC (Uzbek to the USSR), 1400-1500 UTC
15.230	AM		RADIO NATIONAL BARZIL, 0300-0355 UTC
	AM		RADIO MEXICO, 0300-0500 UTC
	AM		RADIO NORWAY INTERNATIONAL, 1600 UTC
15.235	AM		VOICE OF AMERICA, (Service to North Africa), 0730-0800 UTC, (Europe), 1100-1300 UTC, (Central Asia), 1400-1500 UTC, (USSR), 1500-1800 UTC, (Europe), 2100-2300 UTC
	AM		RADIO FREE EUROPE, (Turkmen) 1300-1400 UTC
	AM		RADIO NORWAY INTERNATIONAL, 1900 UTC
15.240	AM		RADIO AUSTRALIA, 0000-0930 UTC
	AM		RADIO MOSCOW, (West Coast), 0530-0900 UTC, (Mar thru Nov)
15.245	AM		VOICE OF AMERICA, (Service to Europe), 1630-1700 UTC, (Middle East), 1800-2000, 2100-2200 UTC

Frequency	Mode	Call Sign	Service / Times
15.250 MHz	AM		VOICE OF AMERICA, (English to Middle East/Europe service) 0100-0200 (Vietnamese to East Asia), 1230-1330 UTC (English to South Asia) 0100-0300, 1400-1500 UTC
15.255	AM		VOICE OF AMERICA, (Service to Middle East), 1330-1500 UTC, (USSR), 1600-1700 UTC, (Europe), 1700-1800 UTC, (USSR), 1800-2100 UTC, (Middle East), 2100-2400 UTC
15.260	AM		BRITISH BROADCASTING CORPORATION, 2000-0330 UTC
	AM		VOICE OF AMERICA, (Service to Europe), 1500-1800 UTC
	AM		ISLAMIC REPUBLIC OF IRAN BROADCASTING, (Spanish Service) 0130-0230, 0530-0630 UTC
15.265	AM		VOICE OF AMERICA, (Spanish to American Republics) 0930-1500 UTC
15.270	AM		VOICE OF AMERICA, (Ukrainian to the USSR), 1600-1800 UTC
	AM	HCJB	THE VOICE OF THE ANDES (Spanish Service) 2200-2230 UTC (English Service) 1900-2000, 2130-2200 UTC

Frequency	Mode	Call Sign	Service / Times
15.280 MHz	AM		RADIO MOSCOW WORLD SERVICE, 1700 UTC
	AM		BRITISH BROADCASTING CORPORATION, 0100-1300 UTC
	AM		VOICE OF AMERICA, (Russian to USSR) 1200-1400 UTC, (Europe), 1600-1900 UTC
15.285	AM		THE VOICE OF QATAR, (English Service) 0245-0700 UTC
15.290	AM		VOICE OF AMERICA, (English to Pacific service) 2200-0100 UTC
	AM		RADIO FREE EUROPE, (Pashto) 0230-0300 UTC, (Dari) 0300-0330 (Russian) 0400-2000 UTC
	AM		RADIO MOSCOW WORLD SERVICE, (English Service) 0000-0300 UTC
15.300	AM		VOICE OF AMERICA, (English to Middle East/Europe service) 1400-2200 UTC
	AM		CHRISTIAN SCIENCE MONITOR, 2200-2255 UTC
15.305	AM		VOICE OF AMERICA, (English to Pacific service) 2200-2400 UTC
15.310	AM		BRITISH BROADCASTING CORPORATION, 0300-0830, 0900-1830 UTC

Frequency	Mode	Call Sign	Service / Times
15.315 MHz	AM		RADIO MOSCOW, (English Service, East Coast) 2330-0300 UTC, (May 5 - July 31)
	AM		VOICE OF AMERICA, (Service to Middle East), 1600-1700 UTC
	AM		BRITISH BROADCASTING CORPORATION, WORLD SERVICE, (Spanish Service), 1300-1330 UTC
	AM		RADIO NETHERLANDS INTERNATIONAL, (Spanish Service), 2330-0025, 0230-0325 UTC
15.320	AM		THE VOICE OF THE UNITED EMIRATES, (English Service) 1330-1400,1600-1700 UTC
	AM		RADIO AUSTRALIA, 2100-0730 UTC
	AM		VOICE OF AMERICA, (Swahili to Africa), 1630-1730 UTC
15.325	AM		THE VOICE OF THE UNITED EMIRATES (English Service) 0530-0600 UTC
	AM		RADIO JAPAN, (English Service) 0300-0330, 1200-1230 UTC
	AM		FRENCH GUIANA, 0330 UTC
	AM		VOICE OF AMERICA, (Russian to USSR) 0800-1100 UTC
15.330	AM		RADIO MOSCOW, 1200-1700 UTC
	AM		VOICE OF AMERICA, (Swahili to Africa), 1630-1730 UTC
	AM		RADIO NETHERLANDS INTERNATIONAL, (Spanish Service), 1200-1225 UTC

Frequency	Mode	Call Sign	Service / Times
15.340 MHz	AM		RADIO FREE EUROPE, 1000-2000 UTC
	AM		BRITISH BROADCASTING CORP., 2000-2300 UTC
15.345	AM		VOICE OF FREE CHINA, TAIWAN, 0200-0400 UTC
	AM		RADIO JAPAN 1700-1800 UTC
15.350	AM		VOICE OF AMERICA, (day break africa, english service) 0300-0700 UTC
15.355	AM		RADIO MOSCOW, (English Service, East Coast) 2300-0200 UTC, (May 5 - Sept 28)
	AM		RADIO MOSCOW WORLD SERVICE, (English Service) 2000-0000 UTC
	AM		RADIO NORWAY INTERNATIONAL, 1500 UTC
	AM	WYFR	FAMILY RADIO NETWORK, (Spanish Service), 1200-1400 UTC (English to Europe), 1600-1700, 1900-2000 UTC
15.360	AM		BRITISH BROADCASTING CORPORATION, 0000-0300, 0600-0930 UTC
	AM		RADIO NORWAY INTERNATIONAL, 0100-0300, 1200-1300, 1600-1700
15.365	AM		RADIO AUSTRALIA, 2200-0700 UTC
15.370	AM		RADIO FREE EUROPE, 1100-2100 UTC
	AM		RADIO FREE ASGHANISTAN, 0230-0330 UTC
15.375	AM		RADIO MOSCOW WORLD SERVICE, (English Service) 1600-2200 UTC
15.380	AM		RADIO FREE EUROPE, (Ukrainian) 1500-2000 UTC
	AM		BRITISH BROADCASTING CORP., 0200-0230, 0300-0330, UTC

Frequency	Mode	Call Sign	Service / Times
15.380 MHz	AM		RADIO EXTERIOR DE ESPANA, (Spanish Service) 0900-1900 UTC
15.390	AM		BRITISH BROADCASTING CORPORATION, WORLD SERVICE, (Spanish Service), 0000-0200, 0300-0430 UTC
15.395	AM		THE VOICE OF QATAR, (English Service) 0245-0700 UTC
	AM		VOICE OF AMERICA, (English to South Asia) 1400-1800 UTC, (China), 0000-0100, 2000-2400 UTC
15.400	AM		THE VOICE OF THE UNITED EMIRATES, (English Service) 1330-1400,1600-1700 UTC
	AM		RADIO MOSCOW, SIMUCASTS ON,15180, 15405, 15425, 15455 AND 16190 (USB) 0530-0730 UTC, (English Service),
	AM		BRITISH BROADCASTING COPR. AFRICAN SERVICE, 0700-1130,1500-2330 UTC
	AM		VOICE OF AMERICA, (Service to Latin America), 0100-0300 UTC
	AM		RADIO FINLAND (English Service) 1230-1430 UTC
15.405	AM		RADIO MOSCOW, SIMUCASTS ON, 15180, 15425, 15455, 15455 AND 16190 (USB) 0530-0730 UTC
	AM		RADIO MOSCOW WORLD SERVICE, (English Service) 1800-2200 UTC
	AM		VOICE OF AMERICA, (English to Middle East/Europe service) 0000-0200,

Frequency	Mode	Call Sign	Service / Times
15.410 MHz	AM		CHRISTIAN SCIENCE MONITOR, 2000-2055 UTC, (Saturday & Sunday), 2205-2355 UTC
15.410	AM		VOICE OF AMERICA, (English to africa serivce) 1600-2200 UTC (Russian to USSR), 0700-1100 UTC
	AM		RADIO MOSCOW, (English Service, East Coast) 2330-0400 UTC, (May 5 - Aug 31) 0530-0800 UTC, (Aug 31 - May 5)
	AM		RADIO MOSCOW WORLD SERVICE, (English Service) 0000-0300, 2200-0000 UTC
	AM		RADIO AUSTRIA INTERNATIONAL, 0530-0600 UTC
15.420	AM		BRITISH BROADCASTING CORPORATION, 1300 UTC
	AM	WRNO	WRNO WORLDWIDE RADIO, 1400-1500, 1500-2300 UTC
15.425	AM		RADIO MOSCOW, (English Service, East Coast) 2300-0400 UTC, (May - Sept 28), 0530-0800 UTC
	AM		RADIO MOSCOW WORLD SERVICE, (English Service) 0000-0300, 1600-0000 UTC
	AM		VOICE OF AMERICA, (English to Pacific service) 1000-1500 UTC
	AM		RADIO DEUTSCHE WELLE, 0300-0500 UTC
15.430	AM		VOICE OF AMERICA, (Service to Central Asia), 0130-0200 UTC
	AM		UNITED NATIONS RADIO, (Spanish service), 2100-2115 UTC
	AM		SWISS RADIO INTERNATIONAL, 0630-0700 UTC

Frequency	Mode	Call Sign	Service / Times
15.435 MHz	AM		THE VOICE OF THE UNITED EMIRATES (English Service) 1330-1400,1600-1700 UTC RADIO MOSCOW WORLD SERVICE, 1700 UTC
	AM		VOICE OF AMERICA, (Urdu to South Asia), 1500-1800 UTC
15.440	AM	WYFR	FAMILY RADIO NETWORK, 0100-0300 UTC
	AM		RADIO BEIJING, CHINA, (English Service), 0900-1100 UTC
15.445	AM		RADIO FREE EUROPE, (Pashto) 1300-1330 UTC, (Dari) 1330-1400
	AM		RADIO CUBA, 0200-0500 UTC
15.450	AM		RADIO AUSTRIA INTERNATIONAL, 0830-0900, 1130-1200, 1330-1400 UTC
	AM		RADIO BEIJING, CHINA, (English Service), 1200-1300 UTC
15.455	AM		RADIO MOSCOW, (English Service, West Coast) 0400-0430 UTC, (May 5 - Aug 31) 0430-0800 UTC, (May 5 - Sept 28)
15.457	VOICE		SPY STATION, SPANISH FEMALE VOICE, SENDS 5-DIGIT TRAFFIC, 0105, 0300 UTC
15.465	AM		RADIO MOSCOW WORLD SERVICE, 1700 UTC
15.485	AM		RADIO MOSCOW WORLD SERVICE, (English Service) 1000-1600, 2000-0000 UTC
15.500	AM		RADIO MOSCOW WORLD SERVICE, (English Service) 1900-2200 UTC
15.502	ARQ		INTERPOL FREQUENCIES

Frequency	Mode	Call Sign	Service / Times
15.505 MHz	AM		RED CROSS BROADCASTING SERVICE, (Mondays), 1040-1057, 1310-1327 UTC
	AM		SWISS RADIO INTERNATIONAL, 1330-1400 UTC
15.520	AM		RADIO MOSCOW WORLD SERVICE, 1700 UTC
15.525	AM		SWISS RADIO INTERNATIONAL, (English Service), 2100-2130 UTC, (Spanish Service), 2130-2200 UTC
15.540	AM		RADIO MOSCOW, 0200-0300 UTC
15.544	USB		SKYKING BROADCASTS
	AM	WYFR	FAMILY RADIO NETWORK, 2145-0245 UTC
15.550	AM		RADIO MOSCOW, 0000-0200 UTC
15.560	AM		RADIO NETHERLANDS, 0100 UTC
15.566	AM	WYFR	FAMILY RADIO NETWORK, 2000-2300 UTC, (English to Europe), 2000-2200 UTC
	ARQ		INTERPOL FREQUENCIES
15.570	AM		SWISS RADIO INTERNATIONAL, 2200-2230 UTC
15.575	AM		RADIO SOUTH KOREA, (KBS), (English Service), 1100-1200,1800-1900, 2030-2130, 2400-0100 UTC, (Spanish Service), 0130-0230, 2215-2300 UTC
15.580	AM		RADIO MOSCOW, (English Service, East Coast) 2300-0400 UTC, (May 5 - Aug 31) 0400-0500 UTC, (May 5 - Aug 31) (West Coast)
	AM		RADIO MOSCOW WORLD SERVICE, (English Service) 1700-2200 UTC

Frequency	Mode	Call Sign	Service / Times
15.580 MHz	AM		VOICE OF AMERICA, (Service to Africa), 1600-2300 UTC
15.590	AM		BRITISH BROADCASTING CORPORATION, 0400-2330 UT
	AM		KOL ISRAEL RUSSIAN SECTION, 2100-2155, 2300-2355 UTC
15.592	ARQ		INTERPOL FREQUENCIES
15.594	ARQ		INTERPOL FREQUENCIES
15.595	AM		RADIO MOSCOW, (English Service, East Coast) 2300-0400 UTC, (May 5 - Aug 31) 0400-0800 UTC, (May 5 - Aug 31) (West Coast)
15.596	CW	KRH51	U.S. DEPT OF STATE, LONDON, GREAT BRITAIN CW AND 50, 75 BAUD RTTY
15.580	AM		VOICE OF AMERICA, (English to Africa service) 1600-2200 UTC
15.600	AM		RADIO BAGHDAD VOICE OF FREE IRAQ, 2230 UTC
15.610	AM		CHRISTIAN SCIENCE MONITOR, 0800-0855, 2000-2055 UTC (Occasional Use Times), 0800-1000, 1200-2200 UTC
15.613	FAX	AXI	DARWIN, AUSTRALIA, WEATHER, 0800-2100 UTC
15.6335	RTTY		NORTH KOREAN NEWS AGENCY
15.640	AM		VOICE OF ISREAL, JERUSALEM, (English service) 0000-0030,0100-0130, 0200-0230, 0500-0515, UTC
	AM		KOL ISRAEL RUSSIAN SECTION, 0500-0540, 2100-2155,2300-2355, UTC
15.644	FAX	VFF	FROBISHER, CANADA
15.650	AM		KOL ISREAL RUSSINA SECTION, 0735-0800, 1200-1230, 1600 UTC
15.647	RTTY		KUWAIT NEWS SERVICE
15.650	AM		VOICE OF AMERICA

Frequency	Mode	Call Sign	Service / Times
15.665 MHz	AM		CHRISTIAN SCIENCE MONITOR, 2000-2400 UTC, (Occasional Use Times), 0800-1000, 1200-1400 UTC, (Spanish Service), (Domingo), 2120-2155 UTC
15.673	AM		VOICE OF AMERICA
15.674	CW		SPY STATION, MORSE CODE, SENDS 5-DIGIT TRAFFIC, 0300 UTC
15.681	VOICE		SPY STATION, ENGLISH FEMALE VOICE, SENDS 5-DIGIT TRAFFIC, 1300, 1400, 1500 UTC
15.684	ARQ		INTERPOL FREQUENCIES
15.738	ARQ		INTERPOL FREQUENCIES
15.752	USB		VOICE OF AMERICA, 0200-0600 UTC (USB)
15.765	AM		VOICE OF AMERICA
15.7675	CW	RJS	ROD DE RJS (Sends 5-digit coded traffic)
15.770	AM		VOICE OF AMERICA
	AM		ICELANDIS NATIONAL BROADCASTING SERVICE, (Icelandic Service) 1410-1440, 1935-2010 UTC
15.790	AM		ICELANDIS NATIONAL BROADCASTING SERVICE, (Icelandic Service) 2300-2335 UTC
15.7774	USB		U.S. ARMY MARS
15.785	FAX		UPI PHOTOS
15.800	AM		VOICE OF AMERICA
15.809	VOICE		SPY STATION, ENGLISH FEMALE VOICE, SENDS 5-DIGIT TRAFFIC, 1200 UTC
15.813	CW	CMU967	RMIZ DE CMU967 QRA 0 QSV QTC ? KKK (Cuban Military)
15.8438	CW	TG	SPY TRANSMISSION 5-DIGIT CODED GROUPS 0400 AND 0430 UTC
15.845	AM		RADIO EGYPT
	VOICE		SPY STATION, SPANISH FEMALE VOICE, SENDS 5-DIGIT TRAFFIC, 0400, 0430 UTC
15.875	AM		VOICE OF AMERICA
15.8775	AM		VOICE OF AMERICA

Frequency	Mode	Call Sign	Service / Times
15.920 MHz	CW	CFH	NAWS DE CFH II ZKR F1 3287 4148.6 6238.6 8306 8306 12441.5 16642 22166 KHZ AR
15.950	FAX	RBI	RUSSIAN WEATHER
15.955	ARQ		INTERPOL FREQUENCIES
15.960	AM		RADIO AUSTRALIA
15.962	USB		SKYKING BROADCASTS
15.971	CW	KKN50	QRA QRA QRA DE KKN50 KKN50 KKN50 QSX 6/10/12/15 K (Will transmit call once every minute)

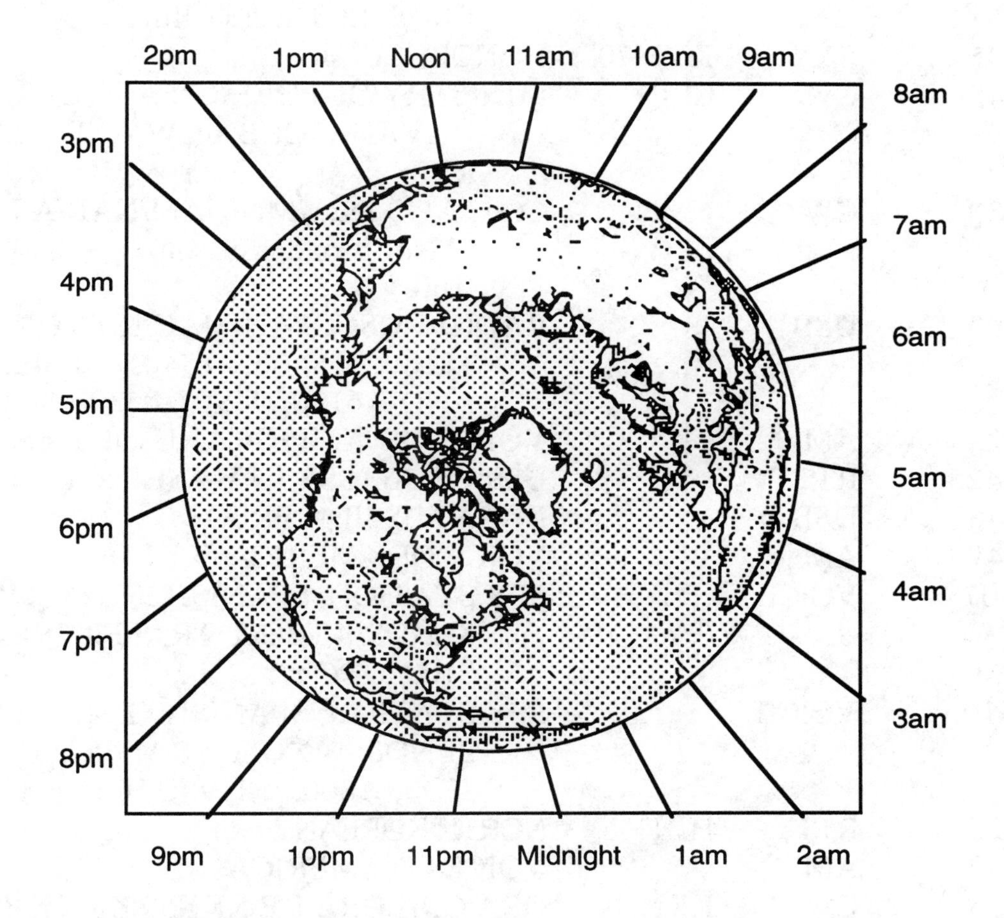

Each "HOUR LINE" is equal to 15 degrees

Frequency	Mode	Call Sign	Service / Times
16.000 MHz	VNG		TIMING SIGNALS FROM AUSTRALIA
16.0285	FAX	BAF9	BEIJING, CHINA
16.085	VOICE		SPY STATION, ENGLISH FEMALE VOICE, SENDS 5-DIGIT TRAFFIC, 1300, 1400, 1500, 1600, 1700, 1730 UTC 3-2 DIGIT TRAFFIC, 1400
16.095	CW	RCX	MINISTRY OF FOREIGN AFFAIRS, USSR, MOSCOW CW AND 50, 75 BAUD RTTY
16.105	CW	CLP1	4CLP DE CLP1 QSA ? QSV K (Cuban), 0550 UTC
16.1085	CW	ONN34	EMBASSY OF REPUBLIC OF IRAN, BRUSSELS, BELGUIM
16.135	FAX	KVM70	HONOLULU, HI
16.145	CW	CLP1	MINISTRY OF FOREIGN AFFAIRS, HAVANA, CUBA CW AND 50 BAUD RTTY
16.180	CW	NMN	CQ CQ CQ DE NMN/NAM/NAR AS SP AR (Also Broadcasts on 05870 at same time)
16.190	USB		RADIO MOSCOW, SIMACASTS ON, 15180, 15400, 15405, 15425 AND 16190 (USB) 0530-0800 UTC
16.222	AM		VOICE OF AMERICA (Used as relay link)
16.2425	RTTY		USSR EMBASSY IN WASHINGTON, D.C.
16.246	USB		NASA MISSION FREQ
16.247	AM		VOICE OF AMERICA
16.307	VOICE		SPY STATION, SPANISH FEMALE VOICE, SENDS 4-DIGIT TRAFFIC, 1830, 2100 UTC
16.310	VOICE		SPY STATION, SPANISH FEMALE VOICE, SENDS 4-DIGIT TRAFFIC, 1700, 1730, 1800, 1839 UTC
16.325	RTTY	HIH	NORTH KOREAN
16.330	AM		VOICE OF AMERICA
16.333	CW	KKN50	QRA QRA QRA DE KKN50 KKN50 KKN50 QSX 6/10/12/16 K (Will transmit call once every minute)
16.338	FAX	ZKLF	WELLINGTON, NEW ZEALAND, WEATHER, 0000-2400 UTC

Frequency	Mode	Call Sign	Service / Times
16.339 MHz	CW	ZKLF	(Sends meteorotogical reports at 0500, 1 800 UTC reports will be in form of 5 digit numbers, FAX at 0545 UTC) Also transmits on 05806 at same time)
16.3401	FAX	ZKLF	NEW ZEALAND
16.3625	CW	KKN50	KWX72 KWX72 KWX72 DE KKN50 KKN50 KKN50 QSX 6/10/12/16 K QRA QRA QRA DE KKN50 KKN50 KKN50 QSX 6/10/12/16 K (Will transmit call once every minute)
16.3866	CW	RIW	SOVIET NAVY
16.410	FAX	NAM	NORFOLK, VA, WEATHER
16.430	CW	CLP1	MINISTRY OF FOREIGN AFFAIRS, HAVANA, CUBA CW AND 50 BAUD RTTY
	VOICE		SPY STATION, ENGLISH MALE VOICE, SENDS 3-2 TRAFFIC, 1300 UTC
16.450	VOICE		SPY STATION, SPANISH FEMALE VOICE, SENDS 5-DIGIT TRAFFIC, 0300 UTC
16.457	CW	NPM	VVV VVV VVV NMO NPM NMO NPM FCM/1 FCM/2 NUKO NUKO NUKO
16.458	CW	KRH50	QRA QRA QRA DE KRH50 KRH50 KRH50 QSX 5/7/11/13/16/20 ? QSX 5/7/11/13/16/20 K (U.S. Embassy, London, England) (Will transmit call once every minute)
16.480	USB		INTERNATIONAL SSB RADIO TELEPHONY (Receive on 17362.00)
16.486	USB		INTERNATIONAL SSB RADIO TELEPHONY (Receive on 17368.00)
16.498	USB		INTERNATIONAL SSB RADIO TELEPHONY (Receive on 17380.00)
16.522	SSB		MARITIME SAFETY AND DISTRESS
16.5375	USB		RAQ MILITARY
16.5871	USB		SHIP TO SHIP, SHIP TO SHORE COMMUNICATIONS.
16.5902	USB		SHIP TO SHIP, SHIP TO SHORE COMMUNICATIONS.
16.5933	USB		SHIP TO SHIP, SHIP TO SHORE COMMUNICATIONS.
16.599	CW	FUF	VVV DE FUF

Frequency	Mode	Call Sign	Service / Times
16.622 MHz	CW	JZEL3	YRF YRF YRF YRF DE JZEL3 JZEL3 JZEL3 JZEL3 GE GE QTC 1 QRV K
16.6655	USB		JAPANESE MARITIME AGENCY
16.7272	USB		JAPANESE MARITIME AGENCY
16.729	CW	GKB	GKB GKB GKB GKB GKB
16.735	CW	XHQ	(Ship calling)
16.7355	CW	P3HJ	NRV NRV NRV DE P3HJ QSA IMI K
	CW	XJYA	JCT JCT's DE YJYA8 QSA NIL UP 625 K
	CW	VAGS	JCT JCT'S DE VAGS K
	CW	9VNW	JNA JNA'S DE 9VNW 9VNW QSA ? QTC K
16.7375	CW	JKPW	JDC JDC's DE JKPW JKPW JKPW PSE QSS QSS QSS 634 II K
	CW	3EZ8	HLJ HLJ HLJ HLJ DE 3EZ8 3EZ8 GE QRU ? QTC 1 K
16.7464	USB		JAPANESE MARITIME AGENCY
16.7683	CW	YUN	CFM DE YUN/LEE KOREA II QSA NIL II K
16.7935	CW	C4FW	KFH DE C4FW MSG QTC QSV K
16.808	ARQ	GKE6	
16.8085	CW	KFS	DE KFS KFS KFS SITOR SELCALL 1.01094 QSX 4 6 8 12 AND 16 MHZ K
16.809	ARQ	WLO	(Receive on 16685.50)
16.811	ARQ	CBV	
	ARQ	VIS	
16.812	ARQ	WLO	(Receive on 16688.50)
16.8125	ARQ	NRV	
16.813	ARQ	VIS	
16.8135	ARQ	KPH	
	ARQ	UAT	
16.814	ARQ	WLO	(Receive on 16695.50)
16.815	ARQ	UPB	
16.8155	ARQ	9MG	
16.816	CW	WNU	DE WNU SELCALL 1109 16816/16692R5
16.817	ARQ	WCC	
	ARQ	NMC	
16.8185	ARQ	KPH	
	ARQ	WLO	(Receive on 16695.50)
	USB		(U.S. COAST GUARD GIVES WEATHER UP-DATES EVERY HALF-HOUR. THIS ANNOUNCEMENT IS MADE IN COMPUTER ELECTRONIC VOICE)

Frequency	Mode	Call Sign	Service / Times
16.8205MHz	ARQ	NMN	
	ARQ	WLO	(Receive on 16607.50)
16.8245	ARQ	WCC	
	ARQ	XSQ	
16.825	ARQ	GKP6	
	ARQ	WCC	
16.826	ARQ	WLO	(Receive on 16703.00)
16.827	ARQ	PCH	
16.8275	ARQ	GKY6	
16.828	ARQ	XSG	
	ARQ	WLO	(Receive on 16705.00)
16.8295	ARQ	UBN	
16.831	ARQ	WLO	(Receive on 16708.00)
16.833	ARQ	WLO	(Receive on 16710.00)
16.8345	ARQ	WNU	DE WNU SELCALL 1109 16834R5/16711R5
16.8365	ARQ	WLO	(Receive on 16713.50)
16.8385	ARQ	PCH	
16.8405	ARQ	GKQ6	
16.8705	CW	KLC	CQ CQ DE KLC KLC QSX 4 6 8 KHZ AND HF 4/6/8/12/16/22 MHZ CHS 5 6 8 KLC TFC LIST ON 1/2 HRS OBS ? AMVER ? QRU ? DE KLC
16.8715	CW	CWA	CQ CQ CQ DE CWA CWA CWA QSX 4/6/8/12/16 MHZ C 5/6/16/18 AND 22 MHZ C 3/4/9 K
16.8767	CW	ZLO	AR DE ZLO ZAY AIA 8 16 22 ZNI 1A J26 8 J76 12 ZNI 1B 12 16 22 AR
16.8809	CW	NMC	CQ CQ CQ DE NMC NMC NMC QRU ? K
16.8815	CW	NOR	CQ CQ CQ DE NOR NOR NOR QRU ? K
16.9055	CW	FUM	VVV DE FUM
16.906	CW	FUV	VVV VVV VVV DE FUV FUV FUV
16.9112	CW	JNA	CQ CQ CQ DE JNA JNA JNA
16.9145	CW	CBV	DE CBV QSX CH 1/3/4
16.916	CW	WSC	CQ CQ DE WSC WSC QSX 6 7 12 16 MHZ OBS ? DE WSC K
16.918	CW	VHP7	VVV VVV VVV DE VHP7 VHP7 VHP7 2/3/4/5/6 AR
16.9185	CW	VIS	VVV VVV VVV DE VIS VIS VIS QSX 2/3/4/5/6 AR
16.9195	CW	VHP	VVV VVV VVV VHP 2/3/4/5/6

Frequency	Mode	Call Sign	Service / Times
16.9275MHz	CW	LGW	CQ CQ DE LGW LGB LGJ LGX LFX QSX 4 CG 8 CG 12 CG 16 CG AND 16740.7 K
16.9332	CW	JOS	CQ CQ CQ DE JOS JOS JOS QSX 16 MHZ
16.9335	CW	WCC	VVV DE WCC QSX 6 8 12 16 22 MHZ OBS AMVER QRU ? K
16.93885	FAX	GYA	NORTHWOOD, UNITED KINGDOM, 0830-1530 UTC
16.9405	CW	XSW	CQ CQ CQ DE XSW XSW XSW QRU ? QSX 500 KHZ AND 8/12/16 MHZ K
16.9476	CW	VIP	VVV DE VIP5 QSX CH 3 4 ET 10
16.9488	CW	VCS	VVV VVV VVV CQ DE VCS VCS VCS QSX 6 8 12 AND 16 MHZ CHNL 3/4/7/8
16.951	CW	6WW	VVV DE 6WW
16.9544	CW	GKC	DE GKC
16.9595	CW	FUM	VVV DE FUM
16.960	CW	FUF	VVV DE FUF
16.9605	CW	CKN	NAWS DE CKN II ZKR F1 2386 4255 6254 8324 12386 22191 MHZ AR
16.9615	CW	FUJ	VVV DE FUJ
16.9645	CW	NMO	VVV VVV VVV DE NMO NPM NMO NPM NMO NPM FCM-1 FCM-2 NUKO NUKO NUKO
16.9675	CW	WLO	CQ CQ DE WLO WLO QSX 434 4257R5 6446R5 8445R5 8473R5 8658R0 12660R0 12704R5 13024R9 16967R5 17172R4 AND 22676R5 KHZ TFC LIST AND WX QSJ 4343 8514 12886R5 17022R5 AND 22487 KHZ
16.968	USB	NMF	USCG, BOSTON, MA
16.9683	CW	WLO	DE WLO OBS ? AMVER ? QSX 4 6 8 12 16 22 25R172 MHZ NW ANS CH/6 K
16.969	CW	PPL	VVV VVV VVV DE PPL PPL PPL QSX K
	USB		U.S. COAST GUARD BOSTON, MA
	FAX	JJC	JAPAN WEATHER
16.9725	CW	WCC	VVV DE WCC WCC WCC BT OBS ? QSX 6 8 12 16 MHZ K

Frequency	Mode	Call Sign	Service / Times
16.9765MHz	CW	NMN	CQ CQ CQ DE NMN NMN NMN QRU ? QSX 8/12/16 MHZ ITU CHANS 4/5/6 BT AS OF 01 JUL 91 BT HERE IS NEW ITU CHANS/FREQS NMN GUARDS BT COMMON CH-4 8369/12553.5/16738 KHZ. CH-5 8367/12551/16735 KHZ. CH-6 8367.5/12551.5/16735.5 KHZ. BT HERE IS THE A1A WORKING BANDS (FREQ CHANS IN .5 KHZ INCREMENTS) 8 MHZ 8342 THRU 8376 KHZ. 12 MKZ 12422 THRU 12476.5 KHZ. 16 MHZ 16619 THRU 16683 KHZ. BT DE NMN QRU ? K
16.9804	CW	DAM	VVV DE DAM DAM
16.984	CW	PPR	VVV DE PPR PPR PPR QSX 22 MHZ
16.9905	CW	HLO	CQ CQ CQ DE HLO HLO HLO QSX 16 MHZ K
16.9985	CW	JDC	CQ CQ CQ DE JDC JDC JDC QSX 16 MHZ K

Frequency	Mode	Call Sign	Service / Times
17.000 MHz	ARQ	WLO	WEATHER/INFO REPORTS 24 HRS-A-DAY
17.002	USB	NMA	USCG, MIAMI, FLA
17.004	CW	HKB	CQ CQ CQ DE HKB HKB HKB QSX ON 8.364/12.546/16.728 MHZ K
17.0045	CW	ZRQ	VVV VVV VVV ZRQ 4/6
17.0054	CW	ZRH	DE ZRH QSX 4 X 8 12 16 X X
17.0065	CW	PCH	DE PCH61 16 K
17.0085	CW	KLB	CQ CQ CQ DE KLB KLB KLB QSX 4 6 8 12 16 AND 22 MHZ OBS ? SITOR SELCAL 1113 AR K
17.0126	CW	HBA	CQ CQ CQ DE HBA HBA HBA
17.0165	CW	K	"K" MARKER
	CW	D	"D" MARKER
	CW	S	"S" MARKER
	CW	C	"C" MARKER
	CW	F	"F" MARKER
	CW	B	"B" MARKER
17.0265	CW	KFS	CQ CQ DE KFS KFS KFS /A QSX 7 12 16 AND 22 MHZ K
17.027	CW	FFL8	CQ CQ CQ DE FFL6 FFL8 FFL6 FFL8 QSX 12 AND 16 MHZ
17.0295	CW	JMC	VVV VVV VVV DE JMC/JMC2/JMC3/JMC5/JMC6
17.0368	CW	XSW	CQ CQ CQ DE XSW XSW XSW QRU ? QSX 500 KHZ AND 8/12/16/22 MHZ K
17.0382	CW	URL	CQ CQ DE URL URL ANS 16668/12468.5 K
17.0385	CW	WNU55	CQ CQ CQ DE WNU55 WNU55 WNU55 QSX 4 6 8 12 16 22 MHZ OBS ?
17.0432	CW	JCU	CQ CQ CQ DE JCU JCU JCU QSX 16 MHZ
17.045	CW	LPD	VVV DE LPD 46/25 LPD 46/25 LPD 46/25 QSX 4 8 AND 12 MHZ CH 6 AND 11 K
17.046	CW	HKC	CQ CQ CQ DE HKC HKC HKC QRU ? QSX 8/12/16 MHZ QSW CH 5/6 K
17.0501	CW	4XZ	VVV DE 4XZ 4XZ BT BT
17.0533	CW	JNA	VVV VVV VVV JNA
17.0542	CW	XSX	CQ CQ CQ DE XSX XSX XSX QSX 8/12/16/22 MHZ K
17.058	FAX	CTU	WEATHER, MONSANTO
17.060	CW	4XO	CQ DE 4XO QSX 8C K
17.064	CW	EAD	DE EAD QRL AS
17.0645	CW	EDZ6	DE EDZ6 QSX 16 MHZ CG AR

Frequency	Mode	Call Sign	Service / Times
17.0685MHz	CW	OXZ	CQ CQ DE OXZ2/OXZ3/OXZ4/OXZ6/OXZ8/ OXZ92 ANS 4 8 12 16 MHZ MSG TFC FER BT M/V REGENTSEA BT BT VEXT BT BT WE LSN 4185.4/8369.6/12553.1/16717.1 AND COMMON FREQS BT II
17.069	FAX	JJC	TOKYO, JAPAN, WEATHER
17.074	CW	LGW	CQ CQ CQ DE LGW LGB LGJ LGX LFX LGG QSX 4 CG 8 CG 12 CD AND 16 CG
17.0793	CW	HLF	CQ CQ CQ DE HLF HLF HLF QSX 16 MHZ K
17.082	CW	JFA	CQ CQ CQ DE JFA JFA JFA K
17.089	CW	KPH	VVV DE KPH QSX 4 6 8 12 16 22 K
17.093	CW	JOR	CQ CQ CQ DE JOR JOR JOR QSX 16 MHZ K
17.0935	CW	AQP	VVV VVV VVV DE AQP4/5/6/7
17.0948	CW	SVA	DE SVA
17.0965	CW	VPS80	CQ CQ DE VPS80 VPS80 QSX CHANNELS 3/4/5/6
17.1038	CW	XSG	CQ CQ CQ DE XSG XSG XSG QRU ? QSX 8 12 16 AND 22 MHZ BK
17.1095	RTTY	UFL	RUSSIAN WEATHER
17.1123	CW	JCS	CQ CQ CQ DE JCS JCS JCS 16 MHZ
17.1136	CW	GKB	DE GKB 4
17.1145	CW	DAN	CQ CQ DE DAN DAN 8 CG 16 CG 22 CG NEW GROUP CHANGE
17.1183	CW	WNU45	CQ CQ CQ DE WNU45 WNU45 WNU45 QSX 4 6 8 12 16 22 MHZ OBS ?
17.1225	CW	XDA	DE XDA (Sends News and Wx information in Spanish)
17.140	FAX	PWZ	WEATHER, RIO de JANEIRO, BRAZIL
17.141	CW	UFN	DE UFN 4 DE UFN 4 DE UFN 4 DE UFN 8 UFN 8 UFN 8 DE UFN 12 DE UFN 12 DE UFN 12 DE UFN 16 DE UFN 16 DE UFN 16 DE UFN 22 DE UFN 22 DE UFN 22 DE UFN 22 (Novorossiysk Radio, USSR)
17.1436	CW	DAN	VVV CQ CQ DE DAN DAN 8 CG 16 CG K
17.144	CW	DAT	CQ CQ CQ DE DAT DAT DAT

17

Frequency	Mode	Call Sign	Service / Times
17.1464MHz	CW	4XO	CQ DE 4XO QSX 12 C K
	CW	NRV	CQ CQ CQ DE NRV NRV NRV QRU ? QSP AMVER QTC QSX 16 MHZ CH 9/12 AND 12 MHZ CH 9/13 QRU ? DE NRV NRV NRV K
17.1465	CW	URL	CQ CQ DE URL URL PSE ANS 124225 K
17.1475	CW	URL	CQ CQ DE URL URL ANS 12454.5/16654.5 PSE K
17.151	FAX	NMC	SAN FRANCISCO, CA, WEATHER
17.1515	USB		U.S. COAST GUARD SAN FRANCISCO, CA
17.1527	CW	URL	CQ CQ DE URL URL QSX 16623.5 K
17.1613	CW	VIS6	VVV DE VIS6 K
17.1625	CW	PPO	VVV DE PPO PPO QSX QAP CHNL 4/5 K AR
17.1655	CW	CLA	CQ CQ CQ DE CLA CLA CLA QSX
17.1665	CW	JCT	CQ CQ CQ DE JCT JCT JCT QSX 16 MHZ K
17.170	CW	PPL	VVV VVV VVV DE PPL PPL PPL QSX K
17.1704	CW	ZLB	DE ZLB 2/4/5/6/7 ZLB 2/4/5/6/7 QSX 4/8/12/16/22 MHZ CHLS 3/4/10 BT
	CW	ZLW	DE ZLW QSX 4/8/12/16/22 MHZ CHNL 3/4/10
17.1715	CW	WLO	DE WLO 1 OBS ? AMVERS ? QSX 8 12 16 22 25R071 MHZ NW ANS CH/6 K
17.1746	CW	VAI	CQ CQ CQ DE VAI VAI VAI QSX 4/8/12/16 MHZ CH 4/5 QSX 22 MHZ CH 3/4 OBS/AMVERS/QRJ/WESTREG ? VAI SITOR SELCALL 1.00581 QRU ? K
17.1853	CW	KFS	CQ DE KFS KFS KFS/B QSX 8 12 16 22 MHZ K
17.1885	CW	SVD	DE SVD
17.195	CW	PPR	VVV DE PPR PPR PPR QSX 22 MHZ K
17.1975	ARQ	KHS	
17.198	ARQ	WNU	
17.1985	ARQ	KHS	
	CW	HGZ	DE HGZ QSX CH 3/4/9
17.199	ARQ	NMC	
	CW	PCH60	DE PCH60 16 AS
17.200	CW	WNU35	CQ CQ CQ DE WNU35 WNU35 WNU35 4 6 8 12 16 22 MHZ OBS ?
	ARQ	VIP	
17.2005	ARQ	HPP	
17.201	ARQ	FFT81	

Frequency	Mode	Call Sign	Service / Times
17.2015MHz	ARQ	KLC	
	ARQ		WEATHER REPORTS
17.202	ARQ	VIP	
17.2025	ARQ	JNA	
17.203	USB		U.S. COAST GUARD, GUAM
17.2035	ARQ	KPH	
	ARQ	NMF	
17.204	ARQ	KPH	
	ARQ	VIP	
17.205	ARQ	JNA	
17.2055	ARQ	KLC	
17.206	ARQ	WPD	
17.2065	ARQ	WNU	
	CW	IAR	VVV DE IAR IAR K 4 8 12 17 22
17.207	ARQ	NMC	
17.2075	ARQ	KPH	
	ARQ	NMC	
17.208	ARQ	KPH	
	ARQ	KLC	
17.2095	ARQ	WLO	
17.210	ARQ	CLA	
	ARQ	NMO	
17.215	ARQ	CLA	
17.2156	CW	VAI	CQ CQ CQ DE VAI VAI VAI QSX 4/6/8/12/16 MHZ CH7/CH8 CH3/CH4 22 MHZ ON REQUEST OBS/AMVER/QRJ? VAI SITOR SELCALL 1.00571 QRU ? K
17.216	ARQ	KPH	
	ARQ	WCC	
17.2161	CW	JDB	CQ CQ CQ DE JDB JDB JDB QSX 16 MHZ K
17.217	ARQ	WLO	(Receive on 16680.00)
17.2183	CW	CLJ	CQ CQ DE CLJ CLJ QSW 435/6403/8609 TFC LIST AS
17.219	ARQ	WLO	(Receive on 16682.00)
17.221	CW	JOU	CQ CQ CQ DE JOU JOU JOU QSX 16 MHZ K
	ARQ	KLC	
17.228	CW	ZLO	DE ZLO ZAY A1A 6 8 12 ZNI 1A 8 12 ZNI 1B 12B 12 16 17B 8 F6 AR AR
17.2308	CW	CWA	CQ CQ CQ DE CWA CWA CWA QSX 4/6/8/12/16/22 MHZ C 3/4/9/10 K

Frequency	Mode	Call Sign	Service / Times
17.231 MHz	CW	KLC	CQ CQ DE KLC KLC QSX 468 KHZ AND HF 4183/6278/8367/12551/ 16736/22281.5 TFC LIST ON 1/2 HRS OBS ? AMVER ? QRU ? DE KLC K
17.2325	CW	HPP	VVV VVV CQ CQ CQ DE HPP HPP HPP QTC ? AMVER ? OBS ? QSW RTG 500 KHZ/8.589/12.699/17.232.4 MHZ CH 5/6/9 AND RTTY12.583.0/12.480.5/ 16.822.5/16.699.5 MHZ
17.2359	USB	JCGT	SYDNEY AUSTRALIA RADIO
17.2393	CW	XSV	CQ CQ CQ DE XSV XSV XSV QRU ? QSX 8 12 AND 16 MHZ BK K
17.2428	USB	WOM	AT&T SERVICES
17.245	CW	KMI	DE KMI AR
17.2607	USB		U.S. NAVY
17.3073	USB	NMC	U.S. COAST GUARD
17.3402	USB		IRAQ NAVY
17.3669	FAX	5YE	WEATHER, NAIROBI, AFRICA
17.390	ARQ		DEPT. OF STATE IN WASHINGTON, D.C.
17.399	VOICE		SPY STATION, ENGLISH FEMALE VOICE, SENDS 3-2 TRAFFIC, 1500 UTC
17.405	FAX	WWD	LAJOLLA, CA, WEATHER
17.4135	CW	KKN39	QRA QRA QRA DE KKN39 KKN39 KKN39 QSX 4/13/17 K
17.414	USB		SKYKING BROADCASTS
17.4146	USB	4SM	SKYKING BROADCASTS
17.4485	ARQ		SAUDI ARABIA EMBASSY IN CANADA
17.485	CW	CLP1	MINISTRY OF FOREIGN AFFAIRS, HAVANA, CUBA CW AND 50 BAUD RTTY
17.512	CW	KRH51	U.S. DEPT OF STATE, LONDON, G.B. CW AND 50, 75 BAUD RTTY
17.525	AM	WWCR	CHRISTAIN RADIO, NASHVILLE, TN, 1500-2400 UTC
17.553	RTTY		IRAN NEWS AGENCY
17.555	AM		CHRISTIAN SCIENCE WORLD SERVICE, 0000-1200, 1600-2400 UTC (Saturaday & Sunday), 0805-1155 UTC, (Spanish Service), 1605-1655 (Sabado, Domingo), 1805-1855 UTC

Frequency	Mode	Call Sign	Service / Times
17.560 MHz	AM		RADIO MOSCOW, 0200-0600 UTC
17.565	AM		RADIO MOSCOW WORLD SERVICE, 1700 UTC
	AM		SWISS RADIO INTERNATIONAL, 0630-0700 UTC
17.575	AM		THE VOICE OF ISREAL (English Service) 0500-0515, 1100-1130, 2230-2300 UTC
	AM		KOL ISRAEL RUSSIAN SECTION, 0735-0800, 1600-1655, 1900-1940 2100-2155 UTC
17.585	FAX	AOK	ROTA, SPAIN, WEATHER
17.590	AM		THE VOICE OF ISREAL (English Service) 1100-1130 UTC
17.595	AM		RADIO MOROCCO, 1400-1600 UTC
17.605	AM		RADIO MOSCOW (English Serivce, East Coast) 2300-0400 UTC (Sept 1 - Sept 28) 0400-0500 UTC (Sept 1 - Sept 28) (West Coast) 0530-0800 UTC (May 5 - Aug 31) (West Coast)
17.6125	AM	WYFR	FAMILY RADIO NETWORK, 1945-2245 UTC
17.617	USB		SKYKING BROADCASTS
17.620	AM		RADIO FRANCE INTERNATIONAL, 1600-2300 UTC
17.630	AM		THE VOICE OF ISRAEL (English Service) 2000-2030 UTC
	AM		RADIO AUSTRALIA, (South Asia Service), 0000-1000, 1300-1330 1500-1800 UTC
17.640	AM		BRITISH BROADCASTING CORPORATION, 0800-1830 UTC
	AM		VOICE OF AMERICA, (Service to Africa), 1600-2300 UTC

Frequency	Mode	Call Sign	Service / Times
17.655 MHz	AM		RADIO MOSCOW, 0400-0600 UTC, (English Service)
17.662	RTTY		USIA SERVICE 75 BAUD RXREV OFF
17.665	AM		RADIO MOSCOW, (English Service), 0000-0600 UTC
17.670	FAX	LQZ68	B. AIRES, ARG, PRESS
	AM		RED CROSS BROADCASTING SERVICE, (English), 0740-0757, (Mondays), 1040-1057 UTC
	AM		RADIO MOSCOW WORLD SERVICE, (English Service) 1300-1900 UTC
17.690	AM		RADIO MOSCOW (English Service, West Coast) 0500-0900 UTC (Sept 1 - Sept 28)
17.695	AM		BRITISH BROADCASTING CORP., 1630-1730 UTC
	AM		RADIO MOSCOW WORLD SERVICE, (English Service) 1600-2100 UTC
17.700	AM		RADIO MOSCOW (English Service, East Coast) 2300-0400 UTC (Sept 1 - Sept 28) 0400-0500 UTC (Sept 1 - Sept 28) (West Coast)
	AM		BRITISH BROADCASTING CORP., 0900-1530 UTC
17.705	AM		RADIO BEIJING, CHINA, 0000-0230 UTC
	AM		VOICE OF AMERICA, (Swahili to Africa), 1630-1730 UTC (Middle East), 0330-0800 UTC
	AM		RADIO HAVANA CUBA, 2000-2300 UTC
	AM		BRITISH BROADCASTING CORP., 0900-1630 UTC

Frequency	Mode	Call Sign	Service / Times
17.710 MHz	AM		VOICE OF AMERICA, (Service to Haiti), 1630-1700 UTC, (Latin America), 1700-1730 UTC
	AM		RADIO BEIJING, CHINA, (English Service), 0900-1000 UTC
17.715	AM		RADIO AUSTRALIA, 0300-0600, 0900-1200, 2130-2400 UTC
	AM		VOICE OF AMERICA, (English to africa service) 0300-0700 UTC
	AM		VOICE OF AMERICA, (Spanish Service), 1200-2200 UTC
	AM		BRITISH BROADCASTING CORP., 2130-2200 UTC
	AM		RADIO EXTERIOR DE ESPANA (Spanish Service) 0900-1900 UTC
17.720	AM		RADIO MOSCOW (English Service, East Coast) 2300-0400 UTC (Sept 1 - Sept 28), (West Coast), 0500-0800 UTC, (Mar thru Nov)
17.725	AM		RADIO CZECHOSLOVAKIA, (English Service), 0730-0800 UTC
17.730	AM		RADIO NORWAY INTERNATIONAL, (English service) 1900 UTC
	AM		VOICE OF AMERICA (Spanish to American Republics) 1200-2200 UTC
	AM		RADIO MOSCOW, 0300-0400 UTC
	AM		SWISS RADIO INTERNATIONAL, (English Service), 0000-0030 UTC, (Spanish Service), 0030-0100, 0230-0300 UTC

Frequency	Mode	Call Sign	Service / Times
17.735 MHz	AM		VOICE OF AMERICA (English to Pacific service) 2100-0100 UTC
	AM		RADIO MOSCOW (English Service, East Coast) 2300-0000 UTC (May 5 - Aug 31)
	AM		RADIO MOSCOW WORLD SERVICE, (English Service) 2100-0000 UTC
	AM		RADIO FREE EUROPE (Kazak) 1100-1400 UTC
17.740	AM		RADIO NORWAY INTERNATIONAL
	AM		RADIO NEW ZEALAND (English Service) 1300-1330 UTC
	AM		VOICE OF AMERICA (English to Middle East/Europe service) 0100-0300 UTC (English to South Asia) 0100-0300 UTC (Swahili to Africa), 1630-1730 UTC
	AM		RADIO SWEDEN, 1300-2030 UTC
	AM		RADIO YUGOSLAVIA, 1200 UTC
17.750	AM		RADIO FREE EUROPE (Tajik) 1100-1200 UTC
	AM		RADIO AUSTRALIA, (South Asia Service), 0000-0400, 0600-1000 UTC
	AM		VOICE OF FREE CHINA, 2200-2300 UTC
	AM		RADIO RED CHINA, 2000-2100 UTC
	AM	WYFR	FAMILY RADIO NETWORK, (English to Europe), 2000-2300

Frequency	Mode	Call Sign	Service / Times
17.755 MHz	AM		VOICE OF AMERICA, (Service to Africa), 1800-1900 UTC
	AM		RADIO NORWAY INTERNATIONAL, 1800 UTC
17.752	CW/RTTY		US EMBASSY, PHILIPPINES
17.760	AM		THE VOICE OF TURKEY (English Service) 0355-0500, 2300-2400 UTC
	AM		RADIO NORWAY INTERNATIONAL, 1700-1900 UTC
	AM		RADIO FREE EUROPE (Tatar-Bashkir) 0900-1000 UTC, (Azerbaijan) 1100-1200 UTC
	AM	WYFR	FAMILY RADIO NETWORK, 1200-1700 UTC
17.765	AM		VOICE OF AMERICA, (Service to China), 2000-0100 UTC
17.770	AM		RADIO NEW ZEALAND INTERNATIONAL, 2206-0630 UTC (Daily)
17.775	AM	KVOH	VOICE OF HOPE (English Service) 1400-0130 UTC
17.780	AM		RADIO MOSCOW WORLD SERVICE, 1700 UTC
	AM		RADIO MOSCOW (English Service, West Coast) 0400-0500 UTC (Sept 1 - Sept 28)
	AM		RADIO FREE EUROPE (Pashto) 1300-1330 UTC, (Dari) 1330-1400
	AM		VOICE OF AMERICA, (Uzbek to the USSR), 1400-1500 UTC
	AM		CHRISTIAN SCIENCE WORLD SERVICE, 0600-0800 UTC (Saturday & Sunday), 0405-0755 UTC
	AM		RADIO NORWAY INTERNATIONAL, 1300-1400 UTC

Frequency	Mode	Call Sign	Service / Times
17.785 MHz	AM		THE VOICE OF TURKEY (English Service) 1330-1400 UTC
	AM		VOICE OF AMERICA, (Service to South Asia), 0030-0230 UTC, (Africa), 1630-1730, 1830-2330 UTC
17.790	AM		RADIO MOSCOW WORLD SERVICE, 1700 UTC
	AM		BRITISH BROADCASTING CORPORATION, 0600-0830, 0900-0930 UTC
	AM	HCJB	QUITO, ECUADOR, 1900-2200 UTC
	AM		VOICE OF AMERICA, (Service to China), 1000-1600 UTC
	AM		RADIO NORWAY INTERNATIONAL, 1500 UTC
	AM		UNITED NATIONS RADIO, (Spanish Service), 1930-2000 UTC
	AM	HCJB	THE VOICE OF THE ANDES (Spanish Service) 2200-2230 UTC (English Service) 1900-2000, 2130-2200 UTC
17.795	AM		RADIO AUSTRALIA, 2130-0700 UTC
	AM		VOICE OF AMERICA (French Service) 0730 UTC
17.800	AM		THE VOICE OF QATAR (English Service) 1700-2130 UTC
	AM		VOICE OF AMERICA (English to Africa service) 1600-2300 UTC
17.805	AM		RADIO FREE EUROPE (Romanian), 0900-1400 UTC

Frequency	Mode	Call Sign	Service / Times
17.810 MHz	AM		VOICE OF AMERICA (English to Middle East/Europe service) 0000-0330 UTC (English to VOA Europe) 0300-0330 UTC (Spanish to American Republics) 0930-1130 UTC
	AM		RADIO JAPAN 0100-0200, 0300-0400, 0500-0600, 0700-0800, 2100-2200, 2300-0000 UTC (Asia)
17.820	AM		VOICE OF AMERICA (English to Pacific Service) 2200-0100 UTC
	AM		RADIO NORWAY INTERNATIONAL, 1200 UTC
17.825	AM		RADIO JAPAN, 0500-0600, 0700-0800 UTC
17.830	AM		THE VOICE OF THE UNITED EMITATES (English Service) 0530-0600 UTC
	AM		BRITISH BROADCASTING CORPORATION, 0600-1030, 1300-2400 UTC
	AM		VOICE OF AMERICA (Spanish to American Republics) 1200-1400 UTC
	AM	WHRI	NOBLESVILLE, IN, 2000-2300 UTC
	AM		RED CROSS BROADCASTING SERVICE, (English), 1310-1327 UTC
	AM		RADIO MOSCOW WORLD SERVICE, (English Service) 1100-1400 UTC
17.835	AM		RADIO FREE EUROPE (Ukrainian) 1500-1700 UTC
	AM		RADIO FREE AFGHANISTAN, 1300-1400 UTC
	AM		RADIO JAPAN 0100-0200 UTC

Frequency	Mode	Call Sign	Service / Times
17.840 MHz	AM		RADIO MOSCOW (English Service, East Coast) 2300-0400 UTC (Sept 1 - Sept 28), (West Coast), 0500-0600 UTC, (Sept thru Nov)
17.845	AM		RADIO JAPAN 0100-0200 UTC
	AM	WFYR	FAMILY RADIO NETWORK, (Spanish Service), 2300-0100 UTC
	AM		RADIO EXTERIOR DE ESPANA, (Spanish Service) 0900-1900 UTC
17.850	AM		RADIO MOSCOW, 0300-0500 UTC
	AM		BRITISH BROADCASTING CORPORATION, WORLD SERVICE, (Spanish Service), 1300-1330 UTC
17.855	AM		VOICE OF AMERICA (Russian to USSR) 1200-1400 UTC (Ukrainian to the USSR), 1600-1800 UTC (Uzbek to the USSR), 1400-1500 UTC
17.860	AM		BRITISH BROADCASTING CORP., 0900-1730 UTC
17.865	AM		VOICE OF AMERICA (Russian to USSR) 0300-1600 UTC
	AM		CHRISTIAN SCIENCE MONITOR, (Sat & Sun), 0005-0355 UTC
17.875	AM		VOICE OF AMERICA, (Urdu to South Asia), 1330-1430 UTC
	AM	HCJB	THE VOICE OF THE ANDES (Spanish Service) 1300-2030 UTC
17.880	AM		BRITISH BROADCASTING CORP., 1400-2130 UTC
	AM		RADIO FREE AFGHANISTAN, 1300-1400 UTC

Frequency	Mode	Call Sign	Service / Times
17.885 MHz	AM		BRITISH BROADCASTING CORPORATION, 0400-0430, 0500-0530, 0600-1400 UTC
	AM		VOICE OF AMERICA (English to Middle East/Europe service) 1400-2400 UTC
17.890	AM		RADIO JAPAN 0500-0600, 0700-0800, 2100-2200 UTC
	AM		VOICE OF AMERICA, (Service to China), 1000-1600 UTC
	AM	HCJB	THE VOICE OF THE ANDES (English Service) 1130-1600 UTC
17.895	AM		VOICE OF AMERICA (English to Middle East/Europe Service) 0200-0330 UTC (English to VOA Europe) 0300-0330 UTC
	AM		RADIO FREE EUROPE (Tajik) 1100-1200 UTC
	AM		RADIO FREE AFGHANISTAN, 0230-0330 UTC
17.925	USB		USED FOR HIJACKING (Emergency Trans by Airline etc.)
17.936	CW		RUSSIAN AIRLINE AEROFLOT
17.940	AM		RADIO BAGHDAD
17.955	AM		RADIO BAGHDAD
17.972	USB		AIR FORCE ONE
	USB		LOOKING GLASS
17.9745	USB		SKYKING BROADCASTS
17.975	USB		USAF SAC AIR/GROUND
	USB		SKYKING BROADCASTS
	USB		LOOKING GLASS
17.985	USB		US NAVY
17.992	USB		LOOKING GLASS, X-RAY 908
17.995	AM		RADIO AUSTRALIA, 0000-0600, 2000-2400 UTC

Frequency	Mode	Call Sign	Service / Times
18.002 MHz	USB		USAF GOBAL COMMAND AND CONTROL
	USB		WEATHER REPORTS FROM McCullen A.F.B.
	USB		SKYKING BROADCASTS
18.005	USB		SKYKING BROADCASTS
18.008	USB		SPACECRAFT, NASA
18.009	USB		NASA MISSION FREQ
	USB		SKYKING BROADCASTS
	USB		USAF TACTICAL COMMAND (Call Signs, Delta 4 Kilo, Bravo 5 Kilo, X-Ray 4 Papa, etc.)
18.019	USB		USAF AIR/GROUND
	USB		USAF TACTICAL AIR COMMAND
	USB		SKYKING BROADCASTS
18.027	USB		USAF TACTICAL USAGE
18.0365	RTTY	JAE	KYODO PRESS, TOKYO NEWS AGENCY, (JAE58/JAT28)
18.046	USB		SKYKING BROADCASTS
18.050	CW	CLP1	MINISTRY OF FOREIGN AFFAIRS, HAVANA, CUBA CW AND 50 BAUD RTTY
18.057	USB		USAF (Used for SAM aircraft)
18.058	FAX	AXI	DARWIN, AUSTRALIA, WEATHER, 0800-2100 UTC
	CW	NPO	U.S. NAVY
18.068	**SSB/CW**	**START**	**OF AMATEUR RADIO 17 METER BAND (Ends 18168.00)**
18.0755	CW	K2G	DE K2G QTK ? QRTK ? K
18.080	FAX	PPN	BRASILIA, BRAZIL
18.087	ARQ		INTERPOL FREQUENCIES
18.093	FAX	LRO	WEATHER, BUENOS AIRES, ARGENTINA
18.097	CW	W1AW	CW BULLETINS
18.110	CW	CLP1	MINISTRY OF FOREIGN AFFAIRS, HAVANA, CUBA CW AND 50 BAUD RTTY
18.1375	AM		VOICE OF AMERICA
18.157	AM		VOICE OF AMERICA
18.160	USB	W1AW	ARRL VOICE BULLETIN
18.16225	RTTY		50 BAUD BAUDOT RXREV OFF

Frequency	Mode	Call Sign	Service / Times
18.185 MHz	CW	CLP1	MINISTRY OF FOREIGN AFFAIRS, HAVANA, CUBA CW AND 50 BAUD RTTY
18.190	ARQ		INTERPOL FREQUENCIES
18.200	CW		SPY STATION, MORSE CODE, SENDS 5-DIGIT TRAFFIC, 1500 UTC
18.210	AM		VOICE OF AMERICA
18.215	AM		VOICE OF AMERICA
18.218	FAX	JMH	TOKYO, JAPAN, WEATHER, 0000-2400 UTC
18.227	FAX	ATP	NEW DELHI, INDIA, 0230-1400 UTC
18.233	FAX	BAF	BEIJING, PEOPLES REPUBLIC OF CHINA, WEATHER
18.25665	RTTY		100 BAUD 6-BIT RTTY RXREV ON
18.265	AM		VOICE OF AMERICA
18.275	AM		VOICE OF AMERICA
18.283	USB/CW		U.S. COAST GUARD
18.292	RTTY	CLN	HAVANA, CUBA NEWS AGENCY
18.297	RTTY		KUWAIT NEWS SERVICE
18.356	VOICE		SPY STATION, ENGLISH FEMALE VOICE, SENDS 5-DIGIT TRAFFIC, 1600 UTC
18.366	RTTY		75 BAUD BAUDOT RXREV ON
18.369	RTTY		75 BAUD BAUDOT RXREV ON
18.380	LSB		100 BAUD ARQ, INTERPOL
18.390	ARQ		INTERPOL FREQUENCIES
18.397	USB		LOOKING GLASS
18.441	FAX	NPN	WEATHER, GUAM
18.450	CW	CLP1	MINISTRY OF FOREIGN AFFAIRS, HAVANA, CUBA CW AND 50 BAUD RTTY
18.515	AM		VOICE OF AMERICA
18.52495	LSB		VOICE OF AMERICA
18.525	CW	KKN50	QRA QRA QRA DE KKN50 KKN50 KKN50 QSX 12/16/18/23 K (Will transmit call once every minute)
	LSB		VOICE OF AMERICA, 0200 UTC (Relay Frequency)
18.55545	ARQ		100 BAUD AMTOR RXREV OFF
18.560	RTTY		IRAN NEWS AGENCY

Frequency	Mode	Call Sign	Service / Times
18.5683MHz	CW	ZWBE	VVV VVV DE ZWBE QSV QSV QRU VVV K AR AR
18.594	USB		SKYKING BROADCASTS
18.607	AM		VOICE OF AMERICA
18.630	CW	CLP1	MINISTRY OF FOREIGN AFFAIRS, HAVANA, CUBA CW AND 50 BAUD RTTY
18.644	RTTY		KUWAIT NEWS SERVICE
18.650	CW	CLP1	MINISTRY OF FOREIGN AFFAIRS, HAVANA, CUBA
18.666	CW		U.S. COAST GUARD
18.755	ARQ		INTERPOL FREQUENCIES
18.756	CW		U.S. COAST GUARD
	ARQ		INTERPOL FREQUENCIES
18.75795	ALIST		100 BAUD ALIST REREV (INTERPOL) 0300 THRU 0315 UTC
18.7825	LSB		VOICE OF AMERICA, 2000 UTC
18.789	USB		INTERNATIONAL SSB RADIO TELEPHONY (Receive on 19773.00)
18.836	RTTY		TASS NEWS AGENCY, MOSCOW, U.S.S.R.
18.865	VOICE		SPY STATION, ENGLISH FEMALE VOICE, SENDS 3-2 DIGIT TRAFFIC, 1500 UTC
18.880	VOICE		SPY STATION, ENGLISH FEMALE VOICE, SENDS 5-DIGIT TRAFFIC, 1500, 1600 UTC
18.887	VOICE		SPY STATION, SPANISH FEMALE VOICE, SENDS 4-DIGIT TRAFFIC, 1630 UTC
18.900	USB	NAV-8	HF/MARS HONOLULU, HI, NAVY MARS
18.922	USB		PLO NET OF EL FATAH
18.9395	FAX	BDF	PEOPLE'S REPUBLIC OF CHINA
18.990	VOICE		SPY STATION, ENGLISH FEMALE VOICE, SENDS 5-DIGIT TRAFFIC, 1600 UTC

Frequency	Mode	Call Sign	Service / Times
19.007 MHz	ARQ		IRAQ EMBASSY BAGHDAD TO ALGERIA
19.042	CW	JCU	VVV DE JCU QSX 22 MHZ K
19.060	CW	CLP1	MINISTRY OF FOREIGN AFFAIRS, HAVANA, CUBA CW AND 50 BAUD RTTY
19.075	AM		VOICE OF AMERICA
19.130	ARQ		INTERPOL FREQUENCIES
19.132	ARQ		INTERPOL FREQUENCIES
19.155	AM		VOICE OF AMERICA
19.160	CW	CLP1	MINISTRY OF FOREIGN AFFAIRS, HAVANA, CUBA CW AND 50 BAUD RTTY
19.200	RTTY		IRAN NEWS AGENCY
19.2033	RTTY		75 BAUD 6-BIT RTTY RXREV OFF
19.20685	TDM		100 BAUD TDM 2:4 RXREV OFF
19.2615	USB		VOICE OF AMERICA, 2000 UTC (Relay Frequency)
19.275	FAX	RXO	RUSSIAN
19.2955	CW	CLP1	VVV VVV VVV CLP7 DE CLP1 QSV QSA IMI K, HAVANA, CUBA
19.303	USB		NASA MISSION FREQ
19.360	LSB		100 BAUD ARQ, INTERPOL
	ARQ		INTERPOL FREQUENCIES
19.363	FAX	NPM	HONOLULU, HAWII
19.405	ARQ		INTERPOL FREQUENCIES
19.420	CW	CLP1	MINISTRY OF FOREIGN AFFAIRS, HAVANA, CUBA CW AND 50 BAUD RTTY
19.445	CW	KRH51	U.S. DEPT OF STATE, LONDON, GREAT BRITAIN CW AND 50, 75 BAUD RTTY
19.46137	ASCII		100 BAUD ASCII RXREV OFF
19.480	AM		VOICE OF AMERICA
19.487	ARQ		INTERPOL FREQUENCIES
19.49697	ALIST		100 BAUD ALIST RXREV OFF
19.501	ARQ	RRG	TASS NEWS SERVICE
19.505	AM		VOICE OF AMERICA
19.52723	RTTY		51 BAUD 6-BIT RTTY RXREV OFF
19.530	AM		VOICE OF AMERICA
	RTTY	JMR	JAPAN WEATHER TRAFFIC
19.5325	USB	WAR	USA MARS, WASHINGTON, D.C.

Frequency	Mode	Call Sign	Service / Times
19.6805MHz	ARQ		196805 to 19703 has not yet been assigned. Receive pairs should 188705 to 18892.00. Frequencies will be assigned to Maritime Ship Telex.
19.688	FAX	AXM	CANBERRA, AUSTRALIA, WEATHER, 0000-2400 UTC
19.715	CW	CLP1	MINISTRY OF FOREIGN AFFAIRS, HAVANA, CUBA CW AND 50 BAUD RTTY
19.7451	BAUDOT		50 BAUD BAUDOT RXREV ON (WX TRAFFIC) CQ CQ CQ DE 6VU23/6VU73/79 RY'S
19.750	FAX	6VU	WEATHER TRAFFIC, (Stations 6VU23/6VU73/6VU79)
19.773	USB	WLO	(Receive on 18798.00)
19.9425	CW	KRH51	U.S. DEPT OF STATE, LONDON, GREAT BRITAIN CW AND 50, 75 BAUD RTTY
19.830	RTTY		TASS NEWS AGENCY, MOSCOW U.S.S.R.
19.8615	FAX	RXO	RUSSIAN
19.881	USB		U.S. AIR FORCE MARS
19.980	RTTY		IRAN NEWS AGENCY
19.995	CW	CLP1	MINISTRY OF FOREIGN AFFAIRS, HAVANA, CUBA CW AND 50 BAUD RTTY

DATE	FREQ.	MODE	TIME	STATION	SIGNAL	COMMENTS	SWL SENT	SWL REC'D

Frequency	Mode	Call Sign	Service / Times
20.000 MHz	VOICE	WWV	INTERNATIONAL STANDARDS TIME FREQUENCY
20.015	CW		U.S. COAST GUARD
	FAX	NAM	NORFOLK, VA.
20.050	CW	CLP1	MINISTRY OF FOREIGN AFFAIRS, HAVANA, CUBA CW AND 50 BAUD RTTY
20.1145	RTTY	CLP1	MINISTRY OF FOREIGN AFFAIRS, HAVANA, CUBA
20.124	USB		LOOKING GLASS, WHISKEY 115
20.148	CW	CLP1	MINISTRY OF FOREIGN AFFAIRS, HAVANA, CUBA CW AND 50 BAUD RTTY
20.167	USB		LOOKING GLASS, WHISKEY 116
20.0785	RTTY	FTU	RY'S DE FTU8B TESTING 1234567890
20.1885	USB		U.S. AIR FORCE MARS
20.190	RTTY		U.S. AIR FORCE MARS
20.198	USB		NASA MISSION FREQ
20.224	CW	NMN	CQ CQ CQ DE NMN/NAM/NRK/GXH/AOK WEATHER QRV
20.348.75	BAUDOT		75 BAUD BAUDOT RXREV OFF
20.365	ARQ		DEPARTMENT OF STATE IN WASHINGTON, D.C.
20.375	USB		U.S. NAVY MARS
20.390	USB		NASA MISSION FREQ
20.440	VOICE		SPY STATION, ENGLISH FEMALE VOICE, 3-2 DIGIT TRAFFIC, 1700 UTC
20.450	CW	CLP1	MINISTRY OF FOREIGN AFFAIRS, HAVANA, CUBA CW AND 50 BAUD RTTY
20.468	FAX		NAVY
20.478	BAUDOT		75 BAUD BAUDOT RXREV ON (NEWS) C/S PXF
20.568	CW	KRH50	QRA QRA QRA DE KRH50 KRH50 KRH50 QSX 5/7/11/13/16/20 ? QSX 5/7/11/13/16/20 K (Will transmit call once every minute)
20.618	FAX	GFE	BRACKNELL, UNITED KINGDOM
20.631	USB		USAF SAC AIR/GROUND
	USB		SKYKING BROADCASTS]
20.650	USB	NAU	USN, CEIBA, PTR
20.688	CW	D3M74	EMBASSY OF USSR, LUANDA, ANGLOA

Frequency	Mode	Call Sign	Service / Times
20.700 MHz	CW	CLP1	MINISTRY OF FOREIGN AFFAIRS, HAVANA, CUBA CW AND 50 BAUD RTTY
20.720	CW	KKN50	U.S. DEPT OF STATE, WASHINGTON, D.C. CW AND 50, 75 BAUD RTTY, ARQ
20.721	ARQ		DEPARTMENT OF STATE IN WASHINGTON, D.C.
20.736	FAX		PRESS PHOTOS
20.737	USB		SKYKING BROADCASTS
20.738	FAX		AP PHOTOS
20.740	USB		SKYKING BROADCASTS
20.805	CW	CLP1	MINISTRY OF FOREIGN AFFAIRS, HAVANA, CUBA CW AND 50 BAUD RTTY
20.812	USB		U.S. ARMY SPECIAL FORCES
20.825	RTTY		TASS NEWS AGENCY, MOSCOW, U.S.S.R.
20.846	USB		SKYKING BROADCASTS
20.870	VOICE		SPY STATION, SPANISH FEMALE VOICE, SENDS 4-DIGIT TRAFFIC, 0000, 0030, 0100 UTC
20.872	RTTY		U.S. AIR FORCE MARS
20.873	USB		CIVIL AIR PATROL
20.885	USB		USAF NORAD HQ.
20.890	USB		U.S. AIR FORCE SAC COMMAND
20.921	RTTY		U.S. ARMY MARS
20.936	USB		U.S. NAVY
20.975	RTTY		U.S. ARMY MARS
20.988	USB		U.S. NAVY MARS
20.991	USB		U.S. AIR FORCE MARS
20.994	USB		U.S. ARMY MARS
20.995	CW	CLP1	MINISTRY OF FOREIGN AFFAIRS, HAVANA, CUBA CW AND 50 BAUD RTTY
20.997	USB		U.S. NAVY MARS
20.9985	USB		U.S. NAVY MARS

Frequency	Mode	Call Sign	Service / Times
21.000 MHz	**SSB/CW**		**START OF AMATEUR RADIO 15 METER BAND (Ends 21450)**
21.077	CW	W1AW	CW BULLETINS
21.095	RTTY		ARRL ARRL BULLETINS
21.098	PACKET		INTERNATIONAL PACKET
21.100	PACKET		AMATEUR RADIO PACKET NETS
21.340	USB		SLOW SCAN TELEVISION
21.345	USB		AMATEUR FAX
21.390	USB	W1AW	ARRL VOICE BULLETIN
21.455	USB	HCJB	QUITO, ECUADOR, 0000-0400, 2000-2300 UTC
	SSB	HCJB	THE VOICE OF THE ANDES (Spanish Service) 2200-2230 UTC (English Service) 0500-1600, 0030-0430,1900-2000, 2130-2200 UTC
21.465	USB		RADIO FOR PEACE INTERNATIONAL, 1800-0330 UTC
	AM		UNITED NATIONS RADIO, 2150-2200, 2345-0000 UTC, (Spanish Service), 1400-1415, 1530-1600 UTC
21.470	AM		BRITISH BROADCASTING CORPORATION, 0430-1730 UTC
21.475	AM		VOICE OF AMERICA (Spanish to American Republics) 1700-1730 UTC (Hindi to South Asia) 0030-0130 UTC
21.480	AM		RADIO MOSCOW, (East Coast), 0000-0200 UTC, (Mar thru Nov), 0300-0500 UTC, (Mar thru Mar)
	AM	HCJB	QUITO, ECUADOR, 1600-2300 UTC
	AM		UNITED NATIONS RADIO, (Spanish Service), 1930-2000 UTC

Frequency	Mode	Call Sign	Service / Times
21.485 MHz	AM		VOICE OF AMERICA (English to Africa service) 2000-2200 UTC (French to Africa) 1830-2000 UTC (Hausa to Africa) 1600-1630 UTC (Portuguese to Africa) 1730-1800 UTC (Swahili to Africa) 1630-1730 UTC
21.490	AM		VOICE OF AMERICA (Spanish to American Republics) 0930-1100, 1200-1400 UTC
	AM		BRITISH BROADCASTING CORPORATION, WORLD SERVICE, (English Service), 1500-1530 UTC, (Spanish Service), 1100-1130 UTC
	AM		RADIO AUSTRIA INTERNATIONAL, 0530-0600, 0830-0900, 1530-1600 UTC
21.500	AM		RADIO FRANCE, 1800-2100 UTC
	AM		FAMILY RADIO, 2000-2200 UTC
	AM		VOICE OF AMERICA, (Service to Middle East), 1330-1500 UTC
	AM	WYFR	FAMILY RADIO NETWORK, (English to Europe), 1700-1900
21.510	AM		RADIO FREE EUROPE (Tajik) 1100-1200 UTC, (Uzbek) 1200-1300 (Pashto) 1300-1330 UTC, (Dari) 1330-1400 UTC
	AM		RADIO FREE AFGHANISTAN, 1300-1400 UTC

Frequency	Mode	Call Sign	Service / Times
21.520 MHz	AM		VOICE OF AMERICA (Pashto to Middle East) 1430-1515 UTC (Urdu to South Asia) 1330-1430 UTC
21.525	AM		RADIO AUSTRALIA, 0200-0730 UTC
	AM	WYFR	FAMILY RADIO NETWORK, 2000-2300 UTC, (English to Europe), 1600-1700 UTC
21.530	AM		RADIO MOSCOW, (World Service), 0100-0400 UTC
21.535	AM		THE VOICE OF QATAR (English Service) 0700-1300 UTC
	AM		VOICE OF AMERICA, (Service to Middle East), 1500-2200 UTC
21.540	AM		VOICE OF AMERICA (English to Middle East/Europe service) 1230-2200 UTC
21.545	AM		CHRISTIAN SCIENCE MONITOR, 1800-2000 UTC
21.550	AM		VOICE OF AMERICA (English to Middle East/Europe service) 0100-0300 (English to South Asia) 0100-0300 UTC
	AM		RADIO FINLAND, (English Service) 1330-1530 UTC
21.555	AM		THE VOICE OF QATAR (English Service) 1300-1700 UTC
	AM		RADIO YUGOSLAVIA, 1200 UTC
21.565	AM		COSTA RICA, RADIO FOR PEASE INTERNATIONAL, 2200 UTC

Frequency	Mode	Call Sign	Service / Times
21.570 MHz	AM		RADIO NEW ZEALAND (English Service) 1300-1330 UTC
	AM		VOICE OF AMERICA (English to Middle East/Europe service) 0800-1100, 1330-2200 UTC (English to VOA Europe) 0800-1000 UTC
21.575	AM		RADIO JAPAN 0700-0800 UTC
21.580	AM		VOICE OF AMERICA (Spanish to American Republics) 1200-1500 UTC
21.585	AM		VOICE OF AMERICA (Mandarin to East Asia) 1100-1300 UTC
21.600	AM		VOICE OF AMERICA, (Service to Africa), 0300-0500 UTC
21.605	AM		THE VOICE OF THE UNITED EMIRATES (English Service) 1330-1400, 1600-1700 UTC
21.610	AM		VOICE OF AMERICA (Spanish to American Republics) 1200-1500 UTC
	AM		RADIO JAPAN 0500-0600, 0900-1000 UTC
21.615	AM		VOICE OF AMERICA (English to Middle East/Europe service) 0800-1300 UTC (English to VOA Europe) 0800-1000 UTC
	AM	WYFR	FAMILY RADIO NETWORK, 1900-2200 UTC, (English to Europe), 1600-1700, 1900-2000 UTC

Frequency	Mode	Call Sign	Service / Times
21.625 MHz	AM		VOICE OF AMERICA (English to Africa service) 1600-2200 UTC (Russian to USSR) 1200-1400 UTC (Uzbek to USSR) 1400-1500 UTC
21.630	AM		SWISS RADIO INTERNATIONAL, 1530-1600 UTC
21.660	AM		BRITISH BROADCASTING CORP., 0700-1730 UTC
21.670	AM		CHRISTIAN SCIENCE MONITOR, 1400-1855 UTC
21.665	AM		RADIO FREE EUROPE (KazaK) 1100-1400 UTC
21.610	AM		RADIO JAPAN (English Service) 0300-0330, 1200-1230 UTC
21.675	AM		THE VOICE OF BAGHDAD (English Service) 1700-1900 UTC
21.685	AM		RADIO FRANCE, 2000 UTC
21.690	AM		RADIO MOSCOW, 0300-0500, 2000-2400 UTC
21.695	AM		RADIO NEW NORWAY INTERNATIONAL (English Service) 1200 UTC
	AM		VOICE OF AMERICA, (Service to Middle East), 1600-2200 UTC
	AM		RED CROSS BROADCASTING SERVICE, (English), 0740-0757 UTC, 1310-1327 UTC
	AM		SWISS RADIO INTERNATIONAL, 1000-1030 UTC

Frequency	Mode	Call Sign	Service / Times
21.705 MHz	AM		RADIO NEW ZEALAND (English Service) 2200 UTC
	AM		RADIO NORWAY INTERNATIONAL, 1400-1500, 2200 UTC
	AM		RADIO CZECHOSLOVAKIA, (English Service), 0730-0800 UTC
21.710	AM		KOL ISRAEL RUSSIAN SECTION, 0735-0800, 1200-1230 UTC
21.715	AM		BRITISH BROADCASTING CORP. 0300-1030 UTC
21.720	AM		VOICE OF FREE CHINA, 2200-2300 UTC
	AM	WYFR	FAMILY RADIO NETWORK, 1645-2300 UTC
	AM		VOICE OF AMERICA, (Service to Middle East), 1330-1600 UTC
	AM		RADIO AUSTRALIA, 1100-1430 UTC
21.730	AM		RADIO NORWAY INTERNATIONAL, 1600 UTC
21.740	AM		RADIO AUSTRALIA, 0000-0900, 2200-0400 UTC
21.745	AM		RADIO FREE EUROPE (Bulgarian), 1300-2100 UTC
	AM		VOICE OF AMERICA, (Service to Latin America), 1700-1730 UTC
21.760	AM		VOICE OF AMERICA (Relay link)
21.770	AM		RADIO FRANCE, 1300-1500 UTC
	AM		RED CROSS BROADCASTING SERVICE, (Mondays), 1040-1057 UTC
	AM		SWISS RADIO INTERNATIONAL, 0630-0700 UTC
21.775	AM		RADIO AUSTRALIA, (South Asia Service), 0300-0930,1030-1200 UTC

Frequency	Mode	Call Sign	Service / Times
21.780 MHz	AM		CHRISTIAN SCIENCE MONITOR, 1800-1915 UTC
21.785	AM		RADIO MOSCOW WORLD SERVICE, 1700 UTC
	ARQ		INTERPOL FREQUENCIES
21.790	AM		THE VOICE OF ISREAL (English Service) 1100-1130 UTC
	AM		RADIO MOSCOW, 0400-0500 UTC
21.800	AM		VOICE OF AMERICA
21.807	ARQ		INTERPOL FREQUENCIES
21.810	AM		BRUSSELS CALLING, (Sunday Only), 1230-1330 UTC
21.807	ARQ		INTERPOL
21.815	USB		SKYKING BROADCASTS
21.825	AM		RADIO AUSTRALIA, 0930-1200 UTC
21.837	FAX	NPM	PEARL HARBOR, HI
21.841	RTTY		TASS NEWS AGENCY, MOSCOW, U.S.S.R.
21.890	CW	CLP1	MINISTRY OF FORIEGN AFFAIRS, HAVANA, CUBA CW AND 50 BAUD RTTY
21.9365	USB		NATIONAL HURRICANE SERVICE
21.980	CW	CLP1	MINISTRY OF FOREIGN AFFAIRS, HAVANA, CUBA CW AND 50 BAUD RTTY

DATE	FREQ.	MODE	TIME	STATION	SIGNAL	COMMENTS	SWL SENT	SWL REC'D

Frequency	Mode	Call Sign	Service / Times
22.09985MHz	USB		CHINA MILITARY CODED VOICE GROUPS
22.108	USB		INTERNATIONAL SSB RADIO TELEPHONY (Receive on 22804.00)
22.123	USB		INTERNATIONAL SSB RADIO TELEPHONY (Receive on 22819.00)
22.124	USB		SHIP TO SHIP, SHIP TO SHORE COMMUNICATIONS.
22.1271	USB		SHIP TO SHIP, SHIP TO SHORE COMMUNICATIONS.
22.1302	USB		SHIP TO SHIP, SHIP TO SHORE COMMUNICATIONS.
22.1333	USB		SHIP TO SHIP, SHIP TO SHORE COMMUNICATIONS.
22.135	USB		INTERNATIONAL SSB RADIO TELEPHONY (Receive on 22831.00)
22.1365	USB		SHIP TO SHIP, SHIP TO SHORE COMMUNICATIONS.
22.233	USB		JAPANESE MARITIME AGENCY
22.2436	USB		JAPANESE MARITIME AGENCY
22.245	USB		JAPANESE MARITIME AGENCY
22.2645	CW	JCU	DE JCU QSA NIL QTC QRU KK
22.2823	CW	HQSK6	HLF HLF HLF DE HQSK6 HQSK6 QSW QSW 271 ? ? UP UP K
22.284	CW	3EAU5	KFH KFH KFH KFH DE 3EAU5 3EAU5 3EAU5 GM K
22.395	CW	CLA	CQ CQ DE CLA CLA QSX C/68364R4/12546R6/16728R8 TX 8573R0/12673R5/16961 QSW CLA 20/31/32/41
22.3195	CW	WLO	DE WLO 1 OBS ? AMVER ? QSX 8 12 16 22 25R071 MHZ NW ANS C9 K
22.3255	FAX	KVM	HONOLULU, HAWAII
22.3795	ARQ	KIP	
22.380	ARQ	VPS	
22.381	ARQ	WLO	(Receive on 22289.00)
22.3815	ARQ	9VG	
22.3835	ARQ	KPH	
	ARQ	NRV	
	ARQ	WLO	(Receive on 22291.50)
22.384	ARQ	VIS	
22.3855	ARQ	WNU	DE WNU SELCAL 1109 22385R5 22293R5

Frequency	Mode	Call Sign	Service / Times
22.3865MHz	ARQ	WCC	
22.3885	ARQ	XSG	
22.3895	ARQ	9VG	
22.390	ARQ	NMO	
22.3955	ARQ	KPH	
	CW	HLF	CQ CQ CQ DE HLF HLF HLF QSX 22 MHZ K
22.403	ARQ	WLO	(Receive on 22311.00)
22.404	ARQ	WLO	(Receive on 22312.00)
22.406	ARQ	WLO	(Receive on 22314.00)
22.407	ARQ	WLO	(Receive on 22315.00)
22.4095	ARQ	UFL	
22.412	ARQ	WLO	(Receive on 22320.00)
22.418	ARQ	WLO	(Receive on 22326.00)
22.41825	CW	JCS	VVV DE JCS
22.429	ARQ	WLO	
22.4582	CW	WNU46	CQ CQ CQ DE WNU46 WNU46 WNU46 QSX 16 22 MHZ OBS ?
22.4587	CW	XSX	CQ CQ CQ DE XSX XSX XSX QSX 8/12/16/22 MHZ K
22.4503	CW	PPO	VVV DE PPO PPO PPO QSX CHANNELS 4/5 ON 22 MHZ QSW 22450 QSX CHANNELS 2/3
22.4615	CW	FUJ	VVV DE FUJ
22.4633	CW	JCU	CQ CQ CQ DE JCU JCU JCU QSX 22 MHZ K
22.465	CW	9MG	CQ CQ DE 9MG2/3/11/12 QSX 9MG2/6697 KHZ ON CH 6/15 9MG3/12678 KHZ ON CH 6/15 9MG11/17172.4 KHZ ON CH 6/15 9MG12/22456 KHZ ON CH 4/9 RTF QSX ON ANY COMMON CHANNELS RTTY ON ARQ MODE SELCALL NO 1.03802 QSO 9MG15 ON 8376.5/8416.5 KHZ 9MG16 ON 16692/16815.5 KHZ VHF ON CH 16 DE 9MG AR K
22.473	CW	CBV	CQ DE CBV QSX CH 1/3/4 ON 8 12 167 22 MHZ AUTOTLX 8 12 16 22 SERIE 9 K
22.4735	CW	VIS42	VVV DE VIS42 K
22.476	CW	NMO	CQ CQ CQ DE NMO NMO NMO QRU AMVER ? QSX 8/12 MHZ AMVER CHNL 5/6/11 ITU 22 MHZ AMVER CHNL 3/4 ITU. QLH 8650/12889.5 KHZ AND 22476 KHZ DE NMO NMO NMO QRU ? K

Frequency	Mode	Call Sign	Service / Times
22.4777MHz	CW	KPH	VVV DE KPH QSX 22 16 12 8 6 AS K
22.485	CW	VHP	VVV VVV VVV DE VHP VHP VHP 2/3/4/5/6/7 K
22.487	FEC	WLO	WEATHER REPORTS
22.491	CW	4XO	CQ DE 4XO QSX 8 C K
22.4958	CW	PPJ	VVV DE PPJ PPJ PPJ QSX CH 4/5 K AND 22 MHZ QSJ 22496 KHZ QSX 2/3 K
22.501	CW	UDN	CQ CQ CQ DE UDN/UFN UDN/UFN QSW 450/4245/8571/12891/17141/22501 TFC LIST
22.515	CW	KFS	CQ DE KFS KFS KFS/B QSX 8 12 16 22 MHZ K
22.5245	CW	JFA	CQ CQ CQ DE JFA JFA JFA K
22.525	CW	JOC	CQ CQ CQ DE JOC JOC JOC QSX 22 MHZ K
22.527	CW		U.S. COAST GUARD
22.528	FAX	NRV	U.S. NAVY
22.532	CW	ZLB	CQ CQ CQ DE ZLB ZLB 2/4/5/6/8 QSX 4/8/12/16 MHZ CHL5/6/17 AND 22 MHZ CHL3/4/9 AR
22.5335	CW	ZLW	DE ZLW ZLW QSX 4 8 12 16 AND 22 MHZ CH 3 4 AND 10 BT
22.536	CW	FUM	VVV DE FUM
22.538	CW	FUF	VVV DE FUF
22.539	CW	KLB	CQ CQ DE KLB KLB QSX 4 6 8 12 16 AND 22 MHZ K
22.541	FAX	NPM	HONOLULU, HAWAII
22.5435	CW	7TF8	CQ CQ CQ DE 7TF8 7TF8 7TF8 QSX 12 MHZ C.5/6/7 TKS K (Algiers Radio, Algiers)
22.5444	CW	FUM	VVV DE FUM
22.5555	CW	LSA	VVV TEST VVV TEST VVV TEST DE LSA
22.5575	CW	KFH	VVV DE KPH KPH KPH QRU ? QSX 4 6 8 12 16 22 MHZ K
22.5615	ARQ	KFS	
22.562	ARQ	WNU	
22.5625	ARQ	KFS	
22.563	ARQ	ZLW	
22.5635	ARQ	WLO	
22.564	ARQ	VIP	
22.5655	ARQ	KLC	
22.566	ARQ	VIP	
	CW	XSW	CQ CQ DE XSW XSW K
22.567	ARQ	NMN	

Frequency	Mode	Call Sign	Service / Times
22.5685MHz	ARQ	WLO	
22.570	ARQ	WNU	DE WNU 1109 22201R5
22.571	ARQ	NMC	
22.5715	ARQ	KPH	
	ARQ	WCC	
22.572	CW		U.S. COAST GUARD
22.573	ARQ	JFA	
22.5775	CW	JNA	CQ CQ CQ DE JNA JNA JNA
22.5785	CW	9VG	CQ DE 9VG59/22578.5 KHZ QSX 22 MHZ CH 1 3 4 6 8 10 QSX 22 MHZ CH 1 3 4 6 8 1 0
22.5875	CW	LDP	VVV DE LPD91/34 LPD91/34 LPD91/34 QSX
22.5895	ARQ	WNU	
	CW	VHI	VVV VVV VVV DE VHI VHI VHI 3/4/5/6/7 AR
22.595	ARQ	KFS	
22.596	ARQ	WOO	
	CW	SVB	DE SVB6/7 QSX 22 CH 5
22.6035	CW	PPR	VVV DE PPR PPR PPR QSX 16 MHZ K
22.6065	CW	JFQ	DE JFQ
22.611	CW	CLA	CQ CQ CQ DE CLA CLA CLA QSX C/6 8368/12552 TX 8573/12673R5 QSW CLA 20/31/32/41 K
22.631	CW	IAR	VVV VVV VVV DE IAR IAR IAR 4 8 12 16 22 AR AR
22.6315	CW	ZLO	DE ZLO ZAY A1A 6 8 12 ZNI 1B 8 12 16 ZNI 1C 4 8 12 AR AR
22.6368	CW	JCT	CQ CQ CQ DE JCT JCT JCT QSX 22 MHZ K
22.638	CW	VAI	CQ CQ CQ DE VAI VAI VAI QSX 4/8/12/16 MHZ CH 4/5 QSX 22 MHZ CH 3/4 OBS/AMVER/QRJ ? VAI SITOR SELCALL 1.00581 QRU? K
22.6467	CW	JOS	CQ CQ CQ DE JOS JOS JOS QSX 22 MHZ K
22.6528	CW	CKN	NAWS DE CKN II ZKR F1 2386 4161 6254 8324 12383 16564 22191 25124 KHZ AR
22.6595	CW	JOR	CQ CQ CQ DE JOR JOR JOR QSX 22 MHZ K
22.6695	CW	PPR	CQ CQ CQ DE PPR PPR PPR QSX 22 MHZ K
22.6702	CW	JCS	CQ CQ CQ DE JCS JCS JCS QSX 22 MHZ K
	CW	CKN	NAWS DE CKN II ZKR F1 2386 4158 6251 8315 12380 16567 22191 KHZ AR
22.671	CW	PPR	VVV DE PPR PPR PPR QSX 16 MHZ K

Frequency	Mode	Call Sign	Service / Times
22.6815MHz	CW	WNU36	CQ CQ CQ DE WNU36 WNU36 WNU36 QSX 16 22 MHZ OBS ?
22.6867	CW	WLO	DE WLO 1 OBS ? AMVERS ? QSX 8 12 16 22 25R172 MHZ NW ANS C 5/6 K
22.6905	CW	VIP06	VVV DE VIP06 QSX CH 3 4 ET 10
22.691	CW	JOU	CQ CQ CQ DE JOU JOU JOU QSX 22 MHZ K
22.693	RTTY	CLA	HAVANA, CUBA, 6-BIT RTTY
22.6942	CW	XSG	CQ CQ CQ DE XSG XSG XSG QRU ? QSX 4 8 12 AND 16 MHZ BT
22.7075	USB		WEATHER INFORMATION
22.740	CW	CLP1	MINISTRY OF FOREIGN AFFAIRS, HAVANA, CUBA CW AND 50 BAUD RTTY
22.755	USB		NASA MISSION FREQ
22.768	FAX	JMH	TOKYO, JAPAN, WEATHER, 0000-2400 UTC
22.781	RTTY		TASS NEWS AGENCY, MOSCOW, U.S.S.R.
22.830	CW	CLP1	MINISTRY OF FOREIGN AFFAIRS, HAVANA, CUBA CW AND 50 BAUD RTTY
22.843	USB		SHIP TO SHIP/SHIP TO SHORE
22.867	FAX	5YE	WEATHER, NAIROBI, AFRICA

Frequency	Mode	Call Sign	Service / Times
23.000 MHz	CW	CLP1	MINISTRY OF FORIEGN AFFAIRS, HAVANA, CUBA CW AND 50 BAUD RTTY
23.010	FAX	NPN	GUAM, MARIANA, ISLAND
23.165	CW	FUV	VVV DE FUV
23.1665	CW	FUM	VVV DE FUM
23.206	USB		USAF TACTICAL AIR COMMAND
	USB		USAF GLOBAL CONTROL AND COMMAND
23.220	USB		USAF FLIGHT WEATHER
23.227	USB		USAF GOBAL CONTROL AND COMMAND
23.330	FAX	KVM	HONOLULU, HI
23.337	USB		USAF SAC AIR/GROUND
	USB		SKYKING BROADCASTS
23.410	LSB		100 BAUD ARQ, INTERPOL
23.419	USB		SKYKING BROADCASTS
23.425	CW	KKN44	QRA QRA QRA DE KKN44 KKN44 KKN44 (Will transmit call once every minute)
23.445	ARQ	KRH51	U.S. EMBASSY LONDON, ENGLAND
23.446	CW	KRH51	U.S. DEPT OF STATE, LONDON, GREAT BRITAIN CW AND 50, 75 BAUD RTTY
23.521	FAX	JMH	FAX FROM JAPAN
23.523	FAX	KVM	HONOLULU, HAWAII
23.635	CW	CLP1	MINISTRY OF FOREIGN AFFAIRS, HAVANA, CUBA CW AND 50 BAUD RTTY
23.640	CW	KWS78	QRA QRA QRA DE KWS78 KWS78 KWS78 QSX 7/10/14/18/23 K (Will transmit call once every minute)
23.860	CW	CLP1	MINISTRY OF FOREIGN AFFAIRS, HAVANA, CUBA CW AND 50 BAUD RTTY
23.772	FEC		UNITED STATES INFORMATION AGENCY
23.9745	RTTY	LOR	ARGENTINIAN NAVY
23.9755	CW	KKN50	QRA QRA QRA DE KKN50 KKN50 KKN50 QSX 12/16/18/23 K (Will transmit call once every minute) CW AND 50, 75 BAUD RTTY

Frequency	Mode	Call Sign	Service / Times
24.0125MHz	USB		U.S. ARMY MARS
24.070	ARQ		INTERPOL FREQUENCIES
24.072	ARQ		INTERPOL FREQUENCIES
24.110	ARQ		INTERPOL FREQUENCIES
24.170	USB		U.S. NAVY MARS
24.301	ARQ		ISRAEL EMBASSY
24.560	USB		U.S. ARMY MARS
24.805	USB	NPG	NAVAL COMMUNICATIONS STATION, STOCKTON, CA
24.806	CW	CLP1	MINISTRY OF FOREIGN AFFAIRS. HAVANA, CUBA CW AND 50 BAUD RTTY
24.810	RTTY	KRH	U.S. EMBASSY LONDON, ENGLAND
24.860	USB		U.S. ARMY MARS
24.890	**CW**	**START OF AMATEUR RADIO 12 METER BAND (Ends 24990.00)**	
24.932	CW	PY2AMI	12 METER BEACON, PY2AMI, GG7IF

TIME CONVERSION CHART

U.T.C.	PST	PDST MST	MDST CST	CDST EST	EDST
0:00	4 pm	5 pm	6 pm	7 pm	8 pm
1:00	5 pm	6 pm	7 pm	8 pm	9 pm
2:00	6 pm	7 pm	8 pm	9 pm	10 pm
3:00	7 pm	8 pm	9 pm	10 pm	11 pm
4:00	8 pm	9 pm	10 pm	11 pm	Midnight
5:00	9 pm	10 pm	11 pm	Midnight	1 am
6:00	10 pm	11 pm	Midnight	1 am	2 am
7:00	11 pm	Midnight	1 am	2 am	3 am
8:00	Midnight	1 am	2 am	3 am	4 am
9:00	1 am	2 am	3 am	4 am	5 am
10:00	2 am	3 am	4 am	5 am	6 am
11:00	3 am	4 am	5 am	6 am	7 am
12:00	4 am	5 am	6 am	7 am	8 am
13:00	5 am	6 am	7 am	8 am	9 am
14:00	6 am	7 am	8 am	9 am	10 am
15:00	7 am	8 am	9 am	10 am	11 am
16:00	8 am	9 am	10 am	11 am	Noon
17:00	9 am	10 am	11 am	Noon	1 pm
18:00	10 am	11 am	Noon	1 pm	2 pm
19:00	11 am	Noon	1 pm	2 pm	3 pm
20:00	Noon	1 pm	2 pm	3 pm	4 pm
21:00	1 pm	2 pm	3 pm	4 pm	5 pm
22:00	2 pm	3 pm	4 pm	5 pm	6 pm
23:00	3 pm	4 pm	5 pm	6 pm	7 pm

Frequency	Mode	Call Sign	Service / Times
25.076 MHz	USB		INTERNATIONAL SSB RADIO TELEPHONY (Receive on 26151.00)
25.196	CW	LIC	DE LIC QSX 22 MHZ AR K
25.210	ARQ	FUB	FRENCH NAVY
25.2225	ARQ	WLO	(Receive on 25075.80)
25.2565	FAX	NPN	U.S. NAVY, GUAM, MARIANAS, ISLAND
25.3475	CW	WLO	DE WLO 1 OBS ? AMVERS ? QSX 8 12 16 22 25R071 MHZ NW ANS CH/6 K
25.3075	CW	LGW	CQ CQ CQ DE LGW LGW LGW
25.3095	CW	LFR	CQ CQ CQ DE LFR LFR LFR
25.378	ARQ	NMC	U.S. COAST GUARD
	CW	NMA	U.S. COAST GUARD
25.5455	CW	KKN39	QRA QRA QRA DE KKN39 KKN39 KKN39 QSX 13/17/25 K (Will transmit call once every minute)
	CW	"R"	"R" MARKER
25.5675	ALST	WLO	WEATHER/SHIP TO SHORE TRAFFIC
25.730	AM		RADIO NORWAY INTERNATIONAL, 1200 AND 1300 UTC
25.750	AM		RADIO AUSTRALIA, 0900-1100 UTC
25.950	USB	HCJB	QUITO, ECUADOR 0200-0300, 1200-1400 UTC

DATE	FREQ.	MODE	TIME	STATION	SIGNAL	COMMENTS	SWL SENT	SWL REC'D

Frequency	Mode	Call Sign	Service / Times
26.030 MHz	FM		REMOTE BROADCASTING
26.070	FM		REMOTE BROADCASTING
26.090	FM		REMOTE BROADCASTING
26.100	FM	ARQ	26.100 TO 26.120 Have yet been assigned. Receive pairs should be 25.173 to 25.192. Frequencies will be assigned to Maritime Ship Telex.
26.1055	ARQ	WLO	(Receive on 25177.50)
26.110	FM		REMOTE BROADCASTING
26.123	CW	WLO	DE WLO 1 OBS ? AMVERS ? QSX 8 12 16 22 25R172 MHZ NW ANS C5/6 K
26.130	FM		REMOTE BROADCASTING
26.150	FM		REMOTE BROADCASTING
26.151	USB	WLO	(Receive on 25076.00)
26.1585	RTTY	UJY	RUSSIAN FISHING SHIP
26.170	CW	L	HF BEACON LOCATED IN LENINGRAD, USSR
26.190	FM		REMOTE BROADCASTING
26.210	FM		REMOTE BROADCASTING
26.230	FM		REMOTE BROADCASTING
26.250	FM		REMOTE BROADCASTING
26.270	FM		REMOTE BROADCASTING
26.290	FM		REMOTE BROADCASTING
26.310	FM		REMOTE BROADCASTING
26.330	FM		REMOTE BROADCASTING
26.350	FM		REMOTE BROADCASTING
26.370	FM		REMOTE BROADCASTING
26.390	FM		REMOTE BROADCASTING
26.410	FM		REMOTE BROADCASTING
26.430	FM		REMOTE BROADCASTING
26.450	FM		REMOTE BROADCASTING
26.470	FM		REMOTE BROADCASTING
26.617	USB		CIVIL AIR PATROL
26.620	USB		CIVIL AIR PATROL
26.670	ARQ		DEPARTMENT OF STATE IN WASHINGTON, D.C.
26.7245	CW	NMN	CQ CQ CQ DE NMN/NAM/NRK/NAR/GXH/ AOK NAVAREA IV QRU
26.965	**AM**		**BEGINING OF CB CHANNELS, ENDS WITH 27405.00, 1 THRU 40**

Frequency	Mode	Call Sign	Service / Times
27.340 MHz	RTTY		MULTI-CHANNEL USAF
27.736	USB		U.S. AIR FORCE MARS
27.74375	CW	Y65	X3, LL1, 0T0 CW MARKING STATIONS
27.710	FM		FOREST PRODUCTS
27.730	FM		FOREST PRODUCTS
27.750	FM		FOREST PRODUCTS
27.770	FM		FOREST PRODUCTS
27.790	FM		FOREST PRODUCTS
27.8295	USB		U.S. AIR FORCE MARS
27.845	LSB		100 BAUD ARQ, INTERPOL
27.870	USB		SKYKING BROADCASTS
27.873	USB		U.S. AIR FORCE
27.900	USB		U.S. NAVY MARS
27.950	USB		U.S. NAVY MARS
27.975	USB		U.S. NAVY MARS
27.995	RTTY		MULTI-CHANNEL USAF
	USB		U.S. ARMY MARS

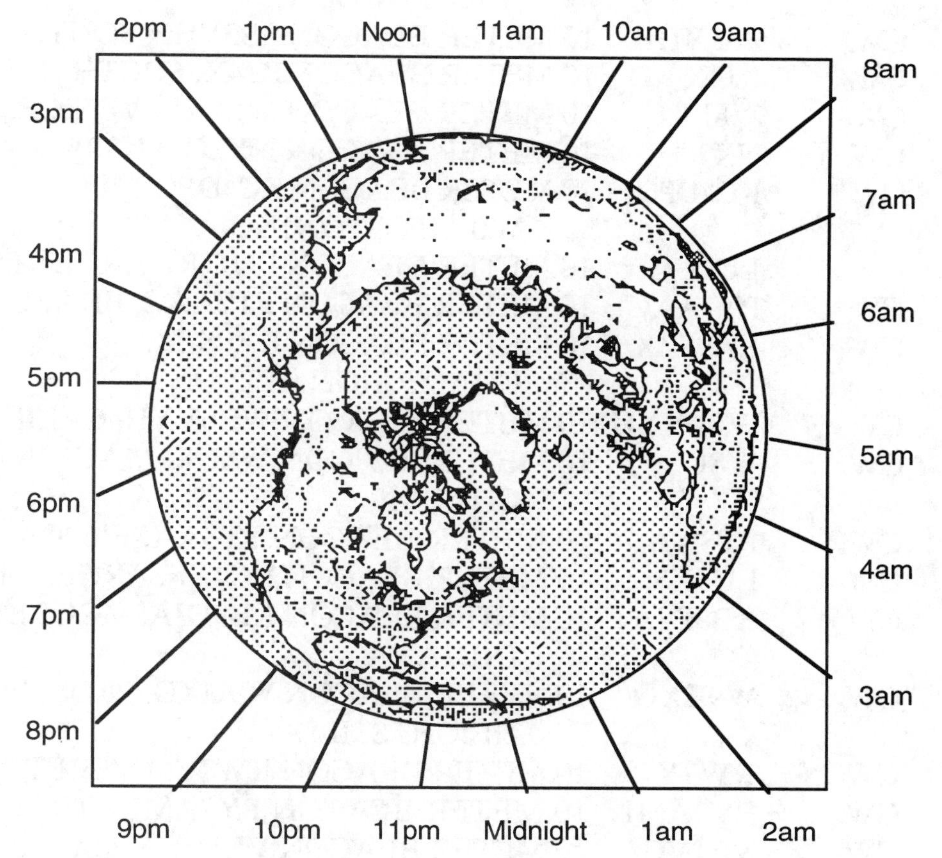

Each "HOUR LINE" is equal to 15 degrees

26
27

Frequency	Mode	Call Sign	Service / Times
28.000MHz	**SSB/CW**		**START OF AMATEUR RADIO 10 METER BAND (Ends 29700.00)**
28.0775	CW	W1AW	CW BULLETINS
28.095	RTTY	ARRL	RTTY BULLETINS
28.100	PACKET		AMATEUR RADIO PACKET NETS
28.1023	PACKET		INTERNATIONAL PACKET
28.113	PACKET	HH2OK	BULLETIN BOARD SYSTEM, HAITI WEST, INDIES
28.150	CW	WA1IOB	10 METER BEACON WA1IOB, MASS.
28.165	CW	VP9AB	10 METER BEACON VP9AB, BERMUDA
28.170	CW	ZL7HMF	10 METER BEACON ZL7HMF, N. ZEALAND
28.175	CW	VE3TEN	10 METER BEACON VE3TEN, CANADA
28.180	CW	ZC4RY	10 METER BEACON ZC4RY, CYPRUS
28.1905	CW	VE6YF	10 METER BEACON VE6YF, CANADA
28.200	CW	JA1GY	10 METER BEACON JA1GY, JAPAN
28.205	CW	WA4SZE	10 METER BEACON WA4SZE/4, HOLLYWOOD, FL
	CW	ZS5VHF	10 METER BEACON ZS5VHF, SOUTH AFRICA
28.2065	CW	KJ4X	10 METER BEACON KJ4X, SOUTH CAROLINA
28.2075	CW	W8FKL	10 METER BEACON W8FKL, VENICE, FLA
28.2095	CW	NX2O	10 METER BEACON NX2O, FN30
28.2105	CW	KC4DPC	10 METER BEACON KC4DPC, WILMINGTON, NC
28.212	CW	LU1UG	10 METER BEACON LU1UG, ARGENTINA
28.213	CW	PT7AAC	10 METER BEACON PT7AAC, BRAZIL
28.216	CW	KA9SZX	10 METER BEACON KA9SZX, CHAMPAIGN, ILL
28.217	CW	WB6VYH	10 METER BEACON WB6VYH, CALIF
28.2185	CW	W8UR	10 METER BEACON W8UR, MACKINAW CITY, MI
28.219	CW	PT8AA	10 METER BEACON PT8AA, FI60CA
28.220	CW	LU4XS	10 METER BEACON LU4XS, ARGENTINA
28.2215	CW	KB9DJA	10 METER BEACON KB9DJA, MOOREVILLE, INDIANA
28.2225	CW	W9UXO	10 METER BEACON W9UXO, PROSPECT HEIGHTS, ILL
28.2255	CW	KW7Y	10 METER BEACON KW7Y, EVERETT, WASH
28.226	CW	PY2AMI	10 METER BEACON PY2AMI, GG67IF
28.227	CW	VE1VCA	10 METER BEACON VE1VCA, CANADA

Frequency	Mode	Call Sign	Service / Times
28.230 MHz	CW	ZL2MHF	10 METER BEACON ZL2MHF, NEW ZEALAND, RE78NU
28.234	CW	KD4EC	10 METER BEACON KD4EC, JUPITER, FLORIDA
28.240	CW	KF9N	10 METER BEACON KF9N/4, TENN 5 WATTS
28.241	CW	KB8JVH	10 METER BEACON KB8JVH, NEWARK, OHIO 20 WATTS
28.2435	CW	WA6APQ	10 METER BEACON WA6APQ, LONG BEACH, CALIF
28.245	CW	ZS1CTB	10 METER BEACON ZS1CTB, PRINCE EDWARD
28.248	CW	EA1AW	10 METER BEACON EA1AW, SPAIN
28.2505	CW	N4MW	10 METER BEACON N4MW, EM55
28.251	CW	VK5WI	10 METER BEACON VK5WI, AUSTRALIA
	CW	K0TFI	10 METER BEACON K0TFI, IOWA
28.252	CW	WJ9Z	10 METER BEACON WJ9Z, WI 55W
28.2585	CW	WB4JHS	10 METER BEACON WB4JHS, KISSIMMEE, FL
28.260	CW	VK5WI	10 METER BEACON VK5WI,M ADELAIDE
28.2615	CW	VK2RSY	10 METER BEACON VK2RSY, LOC QF56MH
28.265	CW	VK6RWA	10 METER BEACON VK6RWA, PERTH
28.270	CW	VK8VF	10 METER BEACON VK8VF, DARWIN PH57
28.271	CW	KF4MS	10 METER BEACON KF4MS, ST. PETERSBURG, FL
28.272	CW	9L1FTN	10 METER BEACON 9L1FTN, SIERRA LEONE
28.2765	CW	NS8V	10 METER BEACON NS8V, GRAND RAPIDS, MICH
28.2785	CW	NØJAR	10 METER BEACON N0JAR, EN31, NEWTON, IOWA
28.281	CW	KG5YB	10 METER BEACON KG5YB, EM22, TYLER, TEXAS
28.2825	CW	VE2HOT	10 METER BEACON VE2HOT, MONTREAL, CANADA, 73 DEG 53 MIN WEST 45 DEG 25 MIN NORTH
28.2835	CW	N2JNT	10 METER BEACON N2JNT, NEW YORK
28.2915	CW	KB9NV	10 METER BEACON KB9NV, COLLINSVILLE, ILL
28.2945	CW	WC8E	10 METER BEACON WC8E, OHIO
28.296	CW	W3VD	10 METER BEACON W3VD, 39R17N, 76R9W
28.295	CW	KEØUL	10 METER BEACON KE0UL, GREELY, CO 5 WATTS
28.298	CW	WA4DJS	10 METER BEACON WA4DJS, FLORIDA, 30 WATTS

Frequency	Mode	Call Sign	Service / Times
28.590 MHz	USB	W1AW	ARRL VOICE BULLETIN
28.680	USB		SLOW SCAN TELEVISION
28.945	USB		AMATEUR FAX
28.888	CW	W6IRK	10 METER BEACON W6IRK, CALIF.
28.992	CW	DFØANN	10 METER BEACON DF0ANN, GERMANY 5 WATTS

TIME CONVERSION CHART

U.T.C.	PST	PDST MST	MDST CST	CDST EST	EDST
0:00	4 pm	5 pm	6 pm	7 pm	8 pm
1:00	5 pm	6 pm	7 pm	8 pm	9 pm
2:00	6 pm	7 pm	8 pm	9 pm	10 pm
3:00	7 pm	8 pm	9 pm	10 pm	11 pm
4:00	8 pm	9 pm	10 pm	11 pm	Midnight
5:00	9 pm	10 pm	11 pm	Midnight	1 am
6:00	10 pm	11 pm	Midnight	1 am	2 am
7:00	11 pm	Midnight	1 am	2 am	3 am
8:00	Midnight	1 am	2 am	3 am	4 am
9:00	1 am	2 am	3 am	4 am	5 am
10:00	2 am	3 am	4 am	5 am	6 am
11:00	3 am	4 am	5 am	6 am	7 am
12:00	4 am	5 am	6 am	7 am	8 am
13:00	5 am	6 am	7 am	8 am	9 am
14:00	6 am	7 am	8 am	9 am	10 am
15:00	7 am	8 am	9 am	10 am	11 am
16:00	8 am	9 am	10 am	11 am	Noon
17:00	9 am	10 am	11 am	Noon	1 pm
18:00	10 am	11 am	Noon	1 pm	2 pm
19:00	11 am	Noon	1 pm	2 pm	3 pm
20:00	Noon	1 pm	2 pm	3 pm	4 pm
21:00	1 pm	2 pm	3 pm	4 pm	5 pm
22:00	2 pm	3 pm	4 pm	5 pm	6 pm
23:00	3 pm	4 pm	5 pm	6 pm	7 pm

Frequency	Mode	Call Sign	Service / Times
29.000 MHz	CW	DLØAR	10 METER BEACON DL0AR, GERMANY
29.1983	ASCII		100 BAUD ASCII CODED MESSAGES RXREV ON
29.300	SSB		29.300 THRU 29.510 MHZ IS DOWN LINK FOR AMATEUR SATELLITE COMMUNICATIONS
29.520	FM		10 METER AMATEUR REPEATER INPUT
29.620	FM		10 METER AMATEUR REPEATER OUTPUT
29.540	FM		10 METER AMATEUR REPEATER INPUT
29.640	FM		10 METER AMATEUR REPEATER OUTPUT
29.560	FM		10 METER AMATEUR REPEATER INPUT
29.600	FM		10 METER FM SIMPLEX OPERATION
29.660	FM		10 METER AMATEUR REPEATER OUTPUT
29.580	FM		10 METER AMATEUR REPEATER INPUT
29.680	FM		10 METER AMATEUR REPEATER OUTPUT
29.875	AM		BRITISH BROADING CORP.
29.8878	RTTY		U.S. AIR FORCE
29.890	DIGITAL		U.S. MILITARY

DATE	FREQ.	MODE	TIME	STATION	SIGNAL	COMMENTS	SWL SENT	SWL REC'D

Amateur Radio Books From 

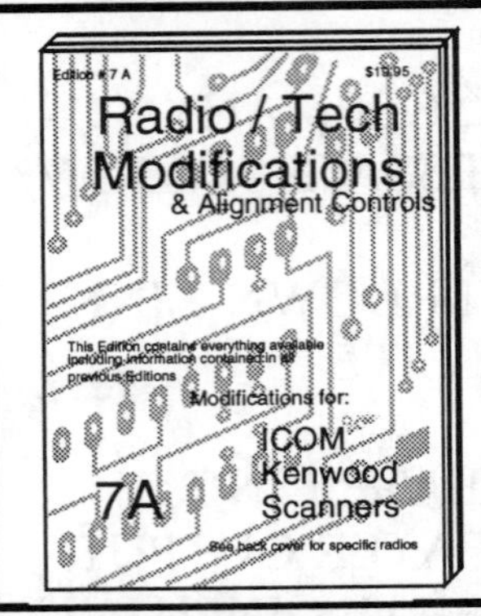

☐ RADIO/TECH MODIFICATIONS 7A & 7B - $19.95 EACH

Volume 7A contains modification information for ICOM, Kenwood, and a wide variety of scanners. Volume 7B contains modification information for Alinco, Standard, Yeasu, CB, and other radio modification information.

Modifications are presented that increase the radio's frequency transmit and reception coverage. Many of the modifications will allow a radio or scanner to monitor cellular phone calls.

The set is updated yearly to include the latest and greatest radios. The timely and accurate information they contain makes it easy for most people to expand their frequency coverage. Previous version are not needed since all available information is included in each edition.

☐ FEDERAL ASSIGNMENTS - $ 24.95

Big brother doesn't stand a chance when the people can tap into the government's secret radio frequencies. Find out what the FBI, CIA or Secret Service is up to. Think the military is up to something, tune in and listen to what's going on.

This book is a communications cornucopia of frequencies for the Federal agencies and the military. 302 pages is organized by both frequency and agency for easy use.

All available government "SECRET" frequencies are listed for the serious listener. Secret service, CIA and FBI are just a few of the US government agencies that are listed in the book.

☐ LOST USER MANUALS - $19.95

Lost the manual for your HT or Mobile rig? Did you purchase a used radio and it did not come with a manual? Do you have the manual but still can't work the radio quickly?

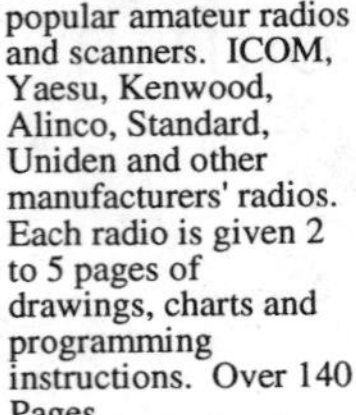

It contains operating instructions for all the popular amateur radios and scanners. ICOM, Yaesu, Kenwood, Alinco, Standard, Uniden and other manufacturers' radios. Each radio is given 2 to 5 pages of drawings, charts and programming instructions. Over 140 Pages

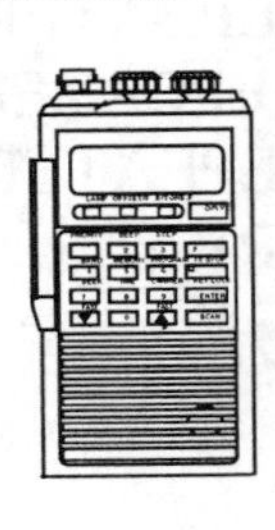

Alinco	Kenwood	ICOM	Yaesu
ALD-24T	TH-21A	IC-02AT	FT-23R
DJ-160T	TH-25A	IC-2A	FT-33R
DJ-560	TH-31A	IC-2GAT	FT-73R
	TH-41A	IC-2SAT	FT-109
Uniden	TH-45A	IC-2SRA	FT-209
BC-100	TH-55A	IC-03AT	FT-211
BC-200	TH-215	IC-3A	FT-212
BC-560	TH-315	IC-3SAT	FT-227R
BC-590	TH-415	IC-04AT	FT-290
BC-760	TM-221	IC-4A	FT-411
	TM-231	IC-4GAT	FT-470
	TM-321	IC-4SAT	FT-709
	TM-421	IC-4SRA	FT-711
	TM-431	IC-12AT	FT-712RH
	TM-521	IC-12GAT	FT-727
	TM-531	IC-24A	FT-811
	TM-621	IC-27A	FT-911
	TM-631	IC-37A	FT-2311
	TM-721	IC-47A	FT-2700
	TM-731	IC-900	FT-4700
	TM-2530	IC-U2AT	
	TM-2550	IC-U4AT	
	TM-2570	IC-W2A	
	TM-3530		
	TR-751		
	TR-851		
	TR-2500		

$ 19.95

U.S. REPEATER MAPBOOK - $ 9.95

Wyoming

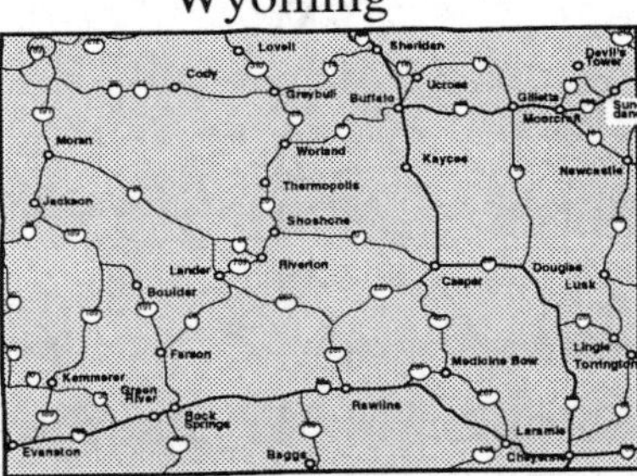

This fifth edition MapBook™ is a perfect travelling companion. The Mapbook contains thousands of open repeaters throughout the U.S.,Canada and Mexico.

NEW DETAILED maps show all highways and major cities in each state. Local repeaters are shown right on the map page in easy to read type. Its never been simplier to find a local frequency while traveling down the road.

NEW table listing of every know open repeater, PL tones and autopatch information are now included.

10, 6, and 2 meter, 220, 440 MHz and 1.2 GHz repeaters.

CD-ROM with Amateur CallSign database & Open Repeaters $ 29.95

Get the U.S. Repeater Mapbook and the CD-ROM for DOS, Windows and MACINTOSH for only $29.95!

☐ HAM RADIO RESOURCE GUIDE FOR SO. CAL. - $ 9.95

Without a doubt the best single information source for amateur radio operators in Southern California. Topics include :

Open repeater guide with unique maps
Autopatch Repeater clubs
Swap Meet & Ham store locations
License exam & class locations
Clubs, & lots of hard-to-find info!

"...an A+ for an extraordinary book. I use it regularly." Gordon West, WB6NOA

☐ AMATEUR HAMBOOK - $ 14.95

Its all here in one handy book. All the technical information, and construction plans needed for amateur radio operation using detailed drawings.

If you have ever wondered how to built an antenna, solder a coax connector or earn a Worked all States (WAS) award its here in simple to read graphic drawings.

- This book will help you keep track of what equipment you have purchased. The handy Equipment log on page 4 will be invaluable if anything is ever stolen.
- Ever wonder what the range of your transmitter is in miles? Page 104
- Installing an N-Connector is simple using the drawing on page 112.
- How many states have you worked on 40 Meters? See page 28.

☐ POLICE AND FIRE COMMUNICATIONS HANDBOOK for So. CALIFORNIA - $19.95

Monitoring the police and fire department has become a popular hobby for hundreds of thousands of people. The mournful wail of an emergency vehicle instantly launches our minds to tales of the tragic and heroic. Tuning in Police and Fire frequencies in Southern California has never been easier than with Artsci's Police and Fire Communications handbook.

This book contains all emergency, government and private industry communications frequencies in Southern California. The frequencies are listed in the most accurate and comprehensive format.

No other publications contain this level of accuracy and comprehensive information. Maps, charts and tables are provided.

☐ RIDING THE AIRWAVES WITH ALPHA & ZULU - $ 14.95

112 cartoon adventures will help anyone from age 8 to 80 learn about Amateur radio. All Novice and Tech Plus question are reviewed and tested.

The cartoon "Phoneticos" are fun to follow through the world of Amateur radio. You'll be ready to pass the FCC test and earn you amateur radio license.

☐ NORTH AMERICAN SHORTWAVE DIRECTORY - $19.95

Tuning in the world on short-wave is easy with these comprehensive and organized lists of broadcast stations. These stations, from most countries of the world, are shown with maps and in tables listed by time and frequency.

There is no easier way to enjoy the endless diversity of the world than by listening to shortwave broadcasts.

The variety of music and entertainment is enormous when you have over 100 countries from which to choose. The differing views of news and opinions opens ones mind to the vast needs and priorities to be found on our ever smaller world. Eight pages of world maps are included that show each countries popular frequencies. The larger type frequency lists makes reference easier for the young and old alike.

Radio/Tech Modifications

The Radio/Tech Modification books are indispensable tools for radio repair technicians and amateur radio hobbyists who are serious about enhancing radio transceivers and scanners.

Modifications are presented that increase the radio's frequency transmit and reception coverage. Many of the modifications will allow a radio or scanner to monitor cellular phone calls.

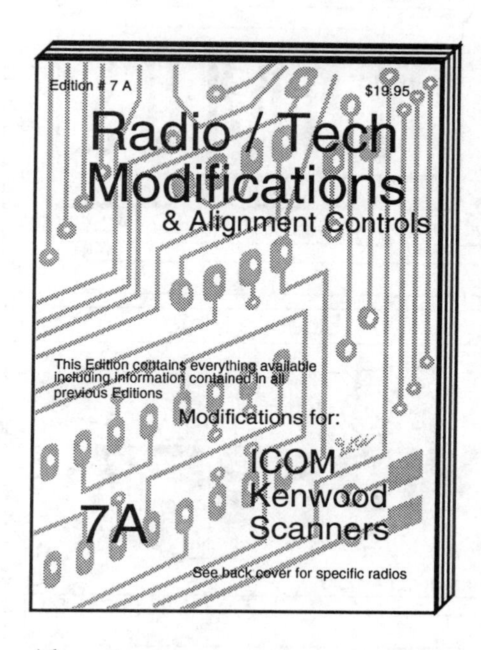

The two volume set contains modification information for all popular handy-talkies, mobile, and base stations. Modifications for a wide variety of scanners and CB radios are also there in concise, easy to follow, instructions. Additionally, alignment controls for many of the radios are presented in graphic line drawings.

Radio/Tech Modifications have been used by Government agencies, Police Departments, and the armed services.

Almost every page contains detailed line drawings to aid the reader in accurately performing the selected modification.

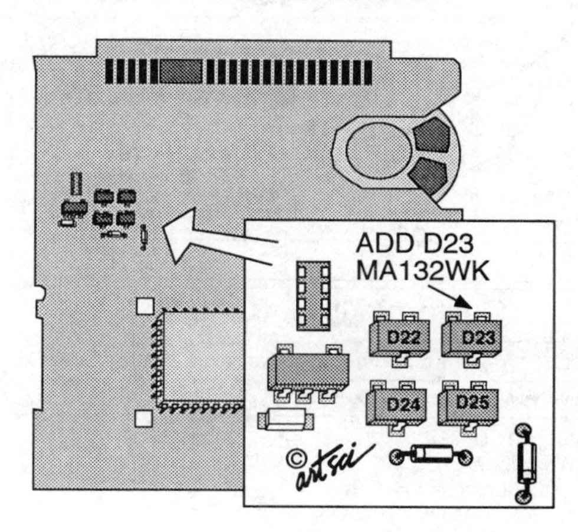

The Radio/Tech Modification Books are updated yearly to include the latest and greatest radios. The timely and accurate information they contain makes it easy for most people to expand their frequency coverage. Each book covers specific brands of radios so you can choose the book you need, or get both so that your library contains powerful information that you, your customers, and your friends can use as needed.

The Radio/Tech Modification Book 7A contains information for ICOM, Kenwood, and a wide variety of scanners. The Radio Tech Mod Book 7B contains Alinco, Standard, Yeasu, CB, and other radio modification information.

QSL Card Order Form

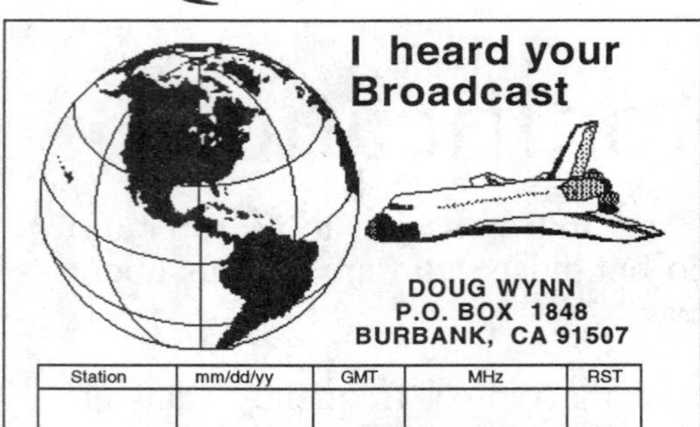

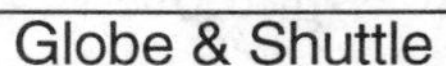

Globe & Shuttle

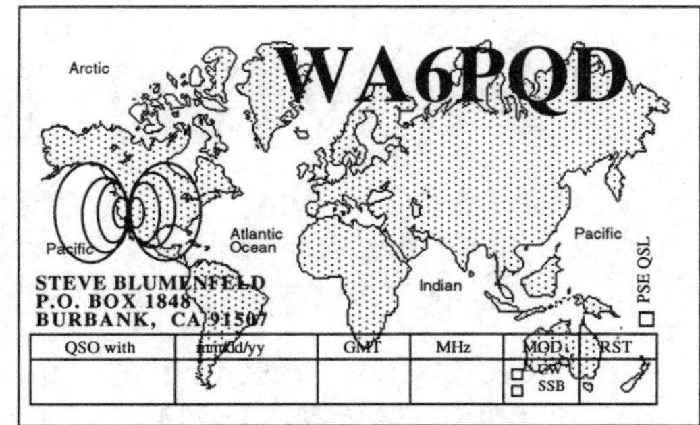

World Map

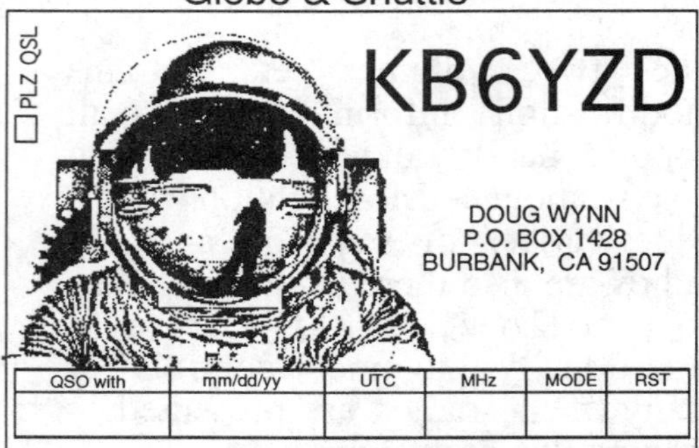

Astronaut

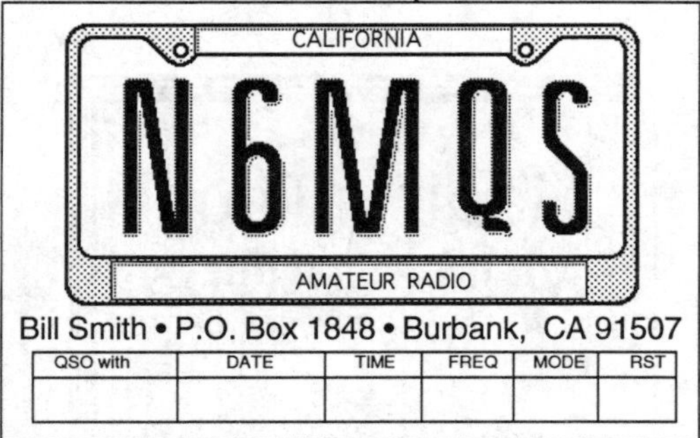

License Plate

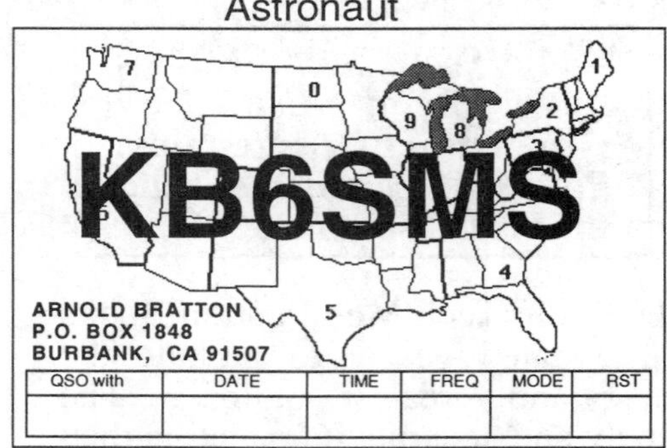

U.S. Map Ham Zones

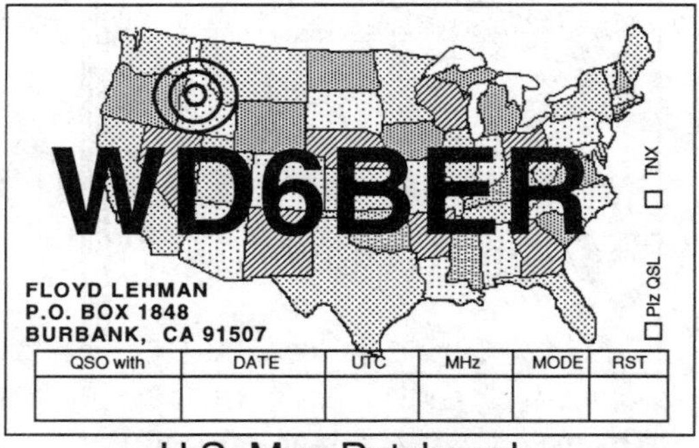

U.S. Map Patchwork

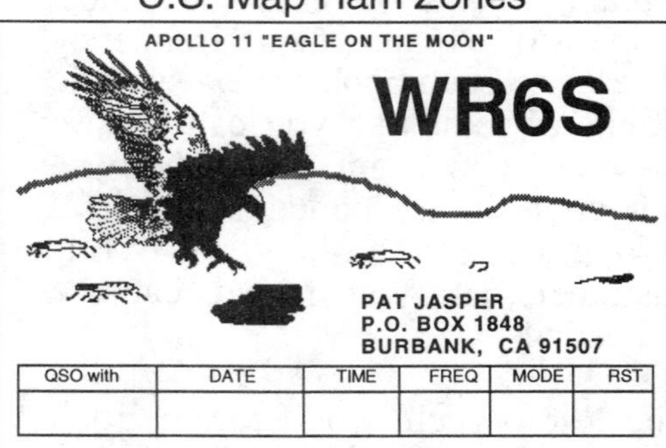

Eagle on the moon

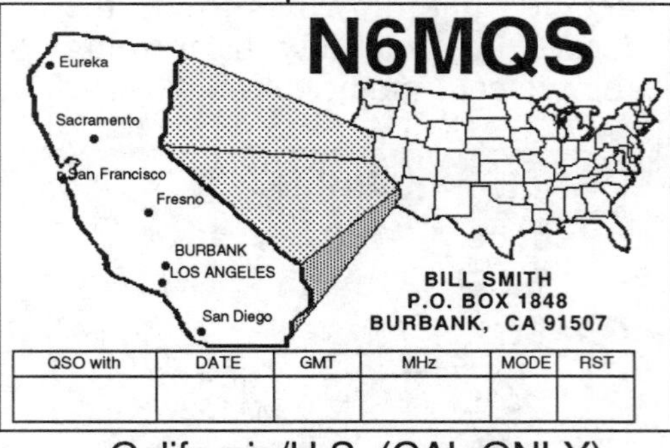

California/U.S. (CAL ONLY)

Call Sign ______________________

Name ______________________

Address ______________________

City ______________________

State Zip ______________________

Phone # (____) ______-________

SELECT desired QSL style.

Delivery will be about 2-3 weeks

All QSL cards are printed on five colors of card stock. This gives you 20 cards in each color.

☐ First 100	$8.95	______
☐ Additional 100's	$5.95	______
☐ 100 Business card size	$7.95	______
☐ Package of QSL Holders	$4.99	______
	Total	______
	TAX8.25%	______
	Shipping	$4.00
	Total Due	______

Mail this form and payment to:

Kamko (818) 843-4080
P.O. Box 1428
Burbank, CA 91507

— ORDER FORM —

	TITLE	DESCRIPTION	PRICE	QTY	EXTENSION
Radio Reference	Radio/Tech Modifications VOL 7A	Over 200 pages of mods for ICOM, KENWOOD Radios & all models of Scanners	19.95		
Radio Reference	Radio/Tech Modifications VOL 7B	Over 200 pages of mods for ALINCO, YAESU, STANDARD and all models of CB equipment.	19.95		
Radio Reference	Lost Users Manuals	Operating Instructions for all popular amateur Mobiles & Ht's.	19.95		
Amateur Reference	U.S. Repeater Mapbook	VHF & UHF Repeater guide for the USA with State Maps showing popular repeaters.	9.95		
Amateur Reference	**CD-ROM** Mapbook Call Sign Database	With CD-ROM- Includes Amateur radio call sign database.	29.95		
Amateur Reference	Amateur HamBook #2	Construction plans, coax, antenna, connector, SWR charts. A must have.	14.95		
Amateur Reference	Ham Radio Resource Guide	For Southern California only. Testing, Club, Repeater, maps & more	9.95		
Scanner/Freq. Reference	Federal Assignments Volume #4	Scanner Frequency guide for all Federal Government Agencies. Over 300 pages	24.95		
Scanner/Freq. Reference	Police & Fire Communications Handbook	For Southern & Central California Scanner Listners. Best available freq. List !!	19.95		
Short-wave	North American Shortwave Directory	Complete Listing of all activity on the HF band 0-30MHz.	19.95		
License Study Guide	Riding the airwaves with Alpha & Zulu	Novice & No-code license test book using cartoon strips to teach. for ages 8 - 80 !!!	19.95		

MAIL ORDER FORM TO:

ARTSCI INC.
P.O. BOX 1428
BURBANK, CA 91507
(818) 843-4080
FAX: (818) 846-2298

SUBTOTAL	
SALES TAX 8.25% CA	
SHIPPING	$ 4.00
ORDER TOTAL	

Shipping charge outside the U.S. is $10.00 or more

SHIP TO:

NAME

ADDRESS

CITY ST ZIP

PHONE ( )

BILLING INFORMATION

CARD #

EXP DATE

Give us your phone number in case we have a problem processing your order.

☐ CHECK ENCLOSED

☐ VISA / MasterCharge / DISCOVER / AMERICAN EXPRESS

artsci